BOGANAIRE
THE RISE AND FALL OF NATHAN TINKLER
PADDY MANNING

Black Inc.

Published by Black Inc.,
an imprint of Schwartz Publishing Pty Ltd
37–39 Langridge Street
Collingwood VIC 3066 Australia
email: enquiries@blackincbooks.com
www.blackincbooks.com

Copyright © Paddy Manning 2014
First published 2013
Paddy Manning asserts his right to be known as the author of this work.

ALL RIGHTS RESERVED.
No part of this publication may be reproduced, stored in a retrieval system, or transmitted in any form by any means electronic, mechanical, photocopying, recording or otherwise without the prior consent of the publishers.

The National Library of Australia Cataloguing-in-Publication entry:

Manning, Paddy, author.

Boganaire : the rise and fall of Nathan Tinkler / Paddy Manning.

9781863956734 (paperback)

Tinkler, Nathan.

Millionaires—Australia—Biography.

Businessmen—Australia—Biography.

Electricians—Australia—Biography.

Mineral industries—Australia.

332.0922

Cover design: Peter Long
Cover photograph: Louie Douvis, Fairfax Photos

CONTENTS

Author's Note		v
1.	Pit Leco: School of Hard Knocks	1
2.	Middlemount: From $1 Million to $442 Million	18
3.	All Too Hard? The Story of Patinack Farm	45
4.	Spendathon: Boys and Their Toys	68
5.	Maules Creek: Lightning Strikes Twice	88
6.	Tinklertown: The White Knight Rides In	118
7.	Whitehaven: The Boardwalk Transaction	160
8.	Judgement Days: Let's Burn a Debt	186
9.	Downfall: Run for Nathan	216
10.	Mulsanne: Billionaire in the Dock	240
11.	Demolition Man: The Briefest Billionaire	259
12.	Epilogue	273
Acknowledgements		284
Notes		287

AUTHOR'S NOTE

This account of Nathan Tinkler's rise and fall was not authorised by him. I tried to speak to Tinkler – as I had tried before, when covering his financial difficulties as a business reporter – but didn't receive any response. As a result, many questions went unanswered, and where there are gaps in this story, it is not for lack of trying, except in one respect: I have not focused my research or inquiries on Nathan Tinkler's private life. In that area, I have by and large relied upon what was already on the public record.

This is the story, then, of an extraordinary young Australian who made and lost an enormous fortune on coal, twice, and who touched many people's lives, for better or worse, particularly through his investments in horses and footy.

Tinkler's story demands telling, although I sometimes wished the job had fallen to somebody else. Just as Tinkler himself wouldn't talk, very many of his creditors and former employees were equally restrained – because they felt intimidated, or had signed a confidentiality agreement, or were still out of pocket, or were considering litigation, or just wanted to move on and put their experiences behind them.

Put simply, almost no one, at any stage, was happy to talk, and very few indeed were willing to go on the record. All the yards have been hard. Nevertheless, I have tried to do Tinkler's remarkable story justice, and to be objective, fair and balanced. Perhaps I felt this responsibility more heavily because neither Tinkler nor his family would cooperate with me. Luckily, however, although Tinkler has generally avoided questioning throughout his career, he has felt a need at times to vent, and he has left more on the record than one might have expected of a supposedly media-shy recluse.

This book, then, is based on a year and a half's reporting and investigation; countless interviews on and off the record; many hours spent reading and rereading media clippings, ASX announcements and company filings (and everything else I could find); and a thousand knockbacks from all and sundry. It shouldn't have been this hard, and this book is dedicated to those people who – because of their sheer honesty, or their courage, or their sense that telling this story was in the public interest – have spoken to me and provided me with information, personal, confidential or otherwise. And to bogans, all of us, everywhere.

1.

PIT LECO
SCHOOL OF HARD KNOCKS

It is the 1980s. Leslie Tinkler's earthmoving business has collapsed and he can no longer afford the rent. He and his young family – wife Zelda, and their two kids, Nathan and Donna – are forced to move into a caravan out the back of his sister's place, at sleepy Port Macquarie, on the mid-north coast of New South Wales.

Accounts vary, but a back injury may have caused Les's misfortune. However it happened, the blow to the Tinkler family was seared into the mind of young Nathan, then at primary school. Decades later, he recalled the experience as the worst of his life, but one which toughened him up. 'These things happen, and when you're a kid something as momentous as that probably lights a fire,' he said. 'I was fortunate to have a very loving family.'

The Tinklers would take years to recover, but recover they did. Starting from next to nothing, with his only qualification an electrical trade certificate, Nathan Tinkler would go on to make not one but two fortunes from brilliantly timed speculations on coal. He would become Australia's youngest ever billionaire at the age of 35, and found a business empire that, at its peak, employed hundreds of people in mining, infrastructure, property, horseracing and football.

When he'd made it, the adult Nathan would remind journalists that he'd grown up 'in a rented house all my life – my parents bought a house when I was fucking 22 years old'. That wasn't exactly right – Les and Zelda bought a house when Tinkler was 17 – but the point was clear: Nathan Tinkler did not grow up with a silver spoon in his mouth. His family expected the worst from the world, and expected to tough it out. As Tinkler told journalist Angus Grigg of the *Australian Financial Review* in the same interview – one of the rare occasions he dropped his guard and talked about his family and his childhood – 'People never want to see you do more than fucking them, I don't give a fuck who they are. The only people who want to see you do better than them is your mum and dad.'

It was true enough. And he did.

*

Nathan Leslie Tinkler was born in Port Macquarie on 1 February 1976, his parents' eldest child. His grandfather, Norman, had been a horse trainer in the Grafton area of northern New South Wales, and a blacksmith, before he moved to the tiny town of Kendall, inland and south from Port Macquarie, tucked in behind a train line and ringed by a state forest. Working as a logging foreman, Norman lived in Kendall for most of his life, and died in August 1979, when Nathan was three. Norman loved his sport, as his obituary in the *Port Macquarie News* recorded:

> He was a respected member of the Camden Haven Urban Committee, foundation member of North Haven Bowling Club and member of its board of directors. He was also actively associated with horse racing and rugby league in Kendall and the Camden Haven area generally.

Nathan would certainly be a chip off the old block.

Les, the younger of Norman's two sons, was born in Port Macquarie in 1950, and had married Zelda in his 20s. The couple moved around a bit – starting in Maitland, then back to Port, then to Inverell, on the Northern Tablelands – as Les's fortunes shifted.

'We certainly were not a wealthy family, not poor, but Nathan did not have anything handed down to him,' Les recalled years later. 'What we have got now we have worked for.'

According to a profile of Nathan Tinkler written by Ben Hills for *Good Weekend*, Tinkler attended a Catholic primary school at Inverell, Holy Trinity. He later went to a number of schools including Inverell High, whose former principal Paul Alliston told Hills that the young Tinkler was an easy-going, friendly sort of lad:

> He liked his sport. He wasn't particularly inclined to study but that didn't make him different from any of the other kids. He needed the odd reminder that school actually was about study and that in between the sport and the social life he probably should knuckle down a little bit more.

Tinkler loved playing footy but, overweight as a kid, he wasn't a standout talent – he freely admitted later that he was generally just 'happy to be picked'. He was nicknamed 'Tink'; when he stuffed up it was 'Tinkerbell'. His former coach Martin Sullivan told Hills that Tinkler was 'not the tallest in the team, but the heaviest' and remembered the day the boy took a tremendous kick to the head: 'When he came out of hospital, someone said, "They don't have that many stitches in a wheat bag." I bet he's still got the scar.' By all accounts a big eater, Tinkler would battle to keep his weight under control into adulthood.

The young Tinkler followed St George and the Brisbane Broncos; at one point he even trialled for the latter. In a 2011 TV interview Tinkler recalled how, as a kid, he idolised the legendary Queensland captain Wally Lewis, 'probably because no one else did! Growing up in NSW … everyone always gave him heaps. But I suppose you've gotta sort of admire the fact that a guy like him always triumphed against sort of adversity.'

More interested in sport than in academic excellence, Tinkler finished in the bottom third of the state for his Higher School Certificate. He recalled later: 'People say what was year 11 and 12 like? I was like, you know, it was a pretty good social couple of years ... won a couple of grand finals ... it was really good! But [at] school, I wasn't too impressive ... getting a trade was a pretty good option for a guy like me.'

A trade wasn't a foregone conclusion, though. His dad had gone into small business as a 'dirt boss', and Tinkler thought at one point of becoming a stock and station agent. When he got an apprenticeship at the local Ford dealer, Les told him: 'You are not fucking doing that. If you don't get out of this town you will never fucking leave.' So in late 1993, at the age of 17, Tinkler got his start at BHP's Bayswater Coal open-cut mine, five kilometres out of Muswellbrook, and began a four-year apprenticeship at the local TAFE as an electrician.

*

Just before Christmas that year, Les and Zelda paid $148,000 for a four-bedroom house on Coolibah Close, Muswellbrook, right near the TAFE where Tinkler would study. The house sat at the top of a hill, with great views of the valley and stunning sunsets over the rough mining town. The Tinklers still own it.

Tinkler joined a local rugby league club, the Muswellbrook Rams, as a reserve-grade journeyman who occasionally started off the bench in its Under-18 side. The club's president, Alan Barker, who played with Tinkler in the early 1990s, remembered him as a pretty good prop forward who could 'cart the ball up the field, like all front rowers do'. The club's historian, Graham Varley, conceded that Tinkler 'needed to be a lot fitter than he was'.

As an apprentice, Tinkler studied one day a week and worked four days. 'Like any apprentice if you aren't a self-starter you quickly learn to be,' Tinkler said later, telling *Master Electrician* magazine that getting his trade was a good foundation that allowed him to achieve his business goals.

Tinkler has described working as a mine electrician – a 'pit leco' – as a 'dirty, mongrelly job' – but he also noted that he 'fucking loved it'. His colleagues from this time have varying memories. Some recall a smart, happy-go-lucky worker. Others reckon he was pretty lazy: he'd sneak the odd nap, and was quick to knock off and head to the pub. According to Hills' profile, Tinkler would jump in a car and drive 50 metres to flick a switch, rather than walk. Rumour has it he was given the nickname 'Standby'. Tinkler had an arrogance about him, some have also recalled, a supreme confidence that he was destined for greater things – that he was going to be a multimillionaire one day. Fellow mine workers remember him reading the newspaper's share pages during his smoko.

While still an apprentice, Tinkler met Rebecca Blackney, said to have worked at a local childcare centre, and soon they were going out. Around this time, too, Tinkler moved out of his parents' place at Coolibah Close and into a rented flat on William Street, a couple of blocks back from the main street and tough old pubs like the Royal, where the lingerie girls are up for a chat.

When Tinkler finished his apprenticeship at the end of 1997, there were no jobs at Bayswater, which later became part of the massive Mount Arthur complex, the biggest thermal coal mine in New South Wales. The next decade's China-fuelled coal boom was not yet on the horizon. Just a few kilometres away, however, building had begun on the Bengalla open-cut mine, and Tinkler was hired by contractor P&H MinePro to work on the dragline.

At the end of 1998, as Bengalla was moving into production, mine owner Peabody started hiring operators for the coal preparation and handling plant. The Howard government had just introduced new industrial relations legislation, and the push was on to replace enterprise bargaining with individual contracts. Rio Tinto had been through a drawn-out dispute in 1997 at its massive Hunter Valley No. 1 mine, which had divided the local mining community.

Years later, Tinkler told Grigg that he had no qualms about working for a non-union company:

> [It] suited me as when I came out of my apprenticeship there was a list of 1,000 union guys who had been laid off. The union rules in those days was that all those guys had to get a job before I was entitled to one. So at the age of 22 I was like 'fuck that, who wants to wait for that?'

Tinkler applied for a job as a technician in the coal handling and preparation plant, sat the interview and got it. A member of the selection panel remembers Tinkler as smart, gregarious, someone who'd be a good fit. Tinkler would work at Bengalla for the next few years, probably pulling about $110,000 to $120,000 a year, including super, in a job that demanded three 12.5-hour shifts each week from four crews working a seven-day roster.

To celebrate his new job, in November 1998 Nathan and Rebecca took the plunge on their first home. They got a mortgage from the Commonwealth Bank and forked out $148,000 for a decent-size home on a good block at 34 Cassidy Avenue, right across the road from the Muswellbrook showground. The couple got married soon afterwards.

After two years at Cassidy Avenue, with their first child on the way – as Tinkler later recalled, 'Up the Valley we didn't get Austar and that back then, so we started early!' – the Tinklers sold up at a small profit and moved to Singleton, buying a house and land package at 6 Clydesdale Close, Hunterview, in March 2001. The land was worth $60,000, and the Tinklers borrowed $252,000 to build a four-bedroom Hotondo Home worth more than $220,000.

They didn't pay their builder, Wayne Kempster. He was offered the Tinklers' job with the requirement that he have the house ready in time for the birth of their first baby. He fulfilled his end of the bargain and had the house ready ten days before Rebecca's due date. In good faith, thinking he was doing the right thing, he let the Tinklers move in before the birth, although they still owed him the final progress payment of $22,000, which was payable within four weeks. It soon became clear to Kempster that the Tinklers had considerably overspent their budget and had not cleared the extra costs with their finance company.

It took Kempster six months to get the final instalment, and extras worth more than $25,000 were still outstanding. He sued the Tinklers, and it would take another two years – a saga of numerous sheriff's orders and rubbery cheques – before he got most of the money, in dribs and drabs. Even then, he finished up more than $2500 short.

When the Tinklers went to sell the home a few years later, the buyer found a defect in the en-suite shower and would not settle unless it was fixed. The Tinklers demanded that Kempster return to fix the problem. As Kempster wrote to the builders' warranty insurer Vero:

> I can not comprehend the gall of these people when I have shown them enormous grace in allowing them the time ... to nearly meet their obligation. On one occasion the Sheriff was two days away from claiming their furniture, until I intervened and called it off. [Tinkler's] father called me and pleaded for more time, going to my local hardware store and putting $1000 on my account against his credit card ... In a further occasion Mr Tinkler asked me in my front yard one afternoon what would I take as a final amount to settle his account. We

agreed on an amount of $5000, which without consulting my accounts seemed short but nevertheless I accepted his personal cheque for this amount. I promptly banked the cheque but it was returned unpaid. There are other occasions of similar events, abusing me over the phone etc etc etc …

Another letter Kempster wrote to Vero recorded that, a year earlier, when he had several sheriff's orders against the Tinklers:

> I put a stay on proceedings against Mr and Mrs Tinkler because the next option was to sell them up. The sheriff had been to the house and indicated the items to be sold to recover the debt owed to me. By that stage they had just had another baby. I thought I was doing the right thing because I was close to having most of my money. They were 'crying poor'. Imagine my anger when I found Mr Tinkler driving a new car when he still owed me thousands of dollars. Mr Tinkler then bought another car for his wife, still owing me thousands of dollars.
>
> I do not believe I am obligated to do any repairs to the house owned by Mr and Mrs Tinkler, currently in the process of being sold. They need to understand that they have to meet their commitments by paying their accounts on time before they can claim any warranties. Should you see it otherwise then once again it would be me being shit on again.

And shat on he was: even though he had not been fully paid, and against the advice of the Master Builders Association, Kempster was forced to go back and fix the defect.

In the middle of his dispute with Kempster, Tinkler lost his job at Bengalla. Rio Tinto's Coal & Allied had taken over Peabody's operations, including Bengalla, in early 2001. Apparently, Tinkler's work ethic and attitude were becoming such a problem that he was moved from shift work to day work so that management could keep a closer eye on him. That meant he would lose his bonuses, and take a pretty hefty pay cut.

And then he was gone altogether. 'He was there one day and gone the next,' recalls one of the men who worked alongside him. At the age of 25, with a mortgage and two young kids, Tinkler's steady job was over. He'd either been sacked – for insubordination, according to one version of the story – or had taken a redundancy. But Bengalla was a brand new mine, just hitting peak production in 2001, so there was no reason anyone should have been made redundant.

Rio Tinto will not discuss the circumstances of his departure. Tinkler told Grigg that he'd left Bengalla out of frustration at 'doing the same shit all the time'. He added: 'I don't do routine and that structured environment didn't agree with me ... there was a communication divide.'

Tinkler wanted a job higher up the mining food chain but realised he wasn't qualified for it. He toyed with the idea of going to university, but, as he later told *BRW* magazine, he felt he would 'fail uni the same way I did high school ... I was just not cut out for hitting the books'.

Tinkler got a job as an electrical field technician for DATEM Moore Technical Services (DMTS), a family-owned contracting business near Singleton; its registered place of business was an unremarkable yard and office at Mount Thorley. At its peak, DMTS had more than 30 employees or sub-contractors. It specialised in providing electrical services to the mining industry. Tinkler got on well with the company's founder, John Moore, and his son Hayden, and was soon promoted to supervisor.

In 2003 Tinkler put in a good word for an electrician he knew of, Matthew Higgins, who began doing field work for DMTS. Higgins had been a mine electrician at Bulga, where he'd also been the union delegate. Xstrata was the mine operator, and after losing a drawn-out dispute with a major contractor to the mine, had been forced to start laying people off. Xstrata had singled out the effective union rep Higgins, making him redundant. The CFMEU filed papers in the Federal Court opposing the redundancy; ultimately, a confidential settlement was paid to Higgins. That money – thought to be in the vicinity of $250,000 – may have been the basis of his subsequent business investment with Tinkler.

On April Fool's Day in 2003, just before the mining boom began, DMTS was put into voluntary administration. In something of a rebirthing exercise, the Moores had decided a month earlier to set up a new company, Support Services Group, run from the same yard in Mount Thorley. Tinkler and Higgins were retained as employees and tried to hold on to as many of DMTS's customers as they could. Hayden Moore was initially a director of the new company, but within months had stepped down and been replaced by Tinkler. Hayden didn't have mining expertise; the business would be Tinkler and Higgins' to run.

Tinkler's own accounts of this period vary, and he has never mentioned the leg-up he got from DMTS. He told Grigg that he had 'started the maintenance business because I really didn't know what to do', and that he had 'no intentions of having it longer term'. But he also told *Master Electrician* magazine that he had:

> … tried to identify a point of difference in the marketplace and for me that was supplying tradesmen that had a high level of experience in the mines as opposed to those that were regularly supplied by labour hire companies. I attracted them through offering full time employment over casual, but it was a tough market. I admire all those young tradesmen that try to take that step from off the floor to being their own boss.

*

According to the Moores, Tinkler had a long-term ambition to develop his own mining venture, and they agreed that Support Services would be the cash cow, generating the money needed to invest in a mining tenement somewhere. The Moore Group would provide engineering, technical and management support.

As he began to evaluate various coal deposits, Tinkler would seek John Moore's advice on technical questions about the equipment needed, the cost structure and so on. The business relationship also extended to property development, with Moore, Tinkler and Higgins at one point proposing to build a small apartment block on a site in Belmont, on the shores of Lake Macquarie. The men's personal relationship was close: the three would go to Knights' rugby league games, lunches and dinners together, often bringing their families. Mysteriously, though, the Australian Securities and Investments Commission (ASIC) has no record of the Moores cancelling their shareholding in Support Services, which would soon change its name to Custom Mining Solutions, with Tinkler and his wife the only shareholders.

Undoubtedly, Tinkler was learning fast. As he told Grigg, he'd always had an interest in the cost structure of mines and had picked up whatever information he could, helped by his experience working as a sparkie on the trucks, draglines and wash plants at Bayswater, which had 'world class machine productivity', and at Bengalla, which had the 'lowest cost profile in the world'.

Tinkler began building his own spreadsheets to describe the cost structure of a mining operation. At some point he got hold of an actual mine financial model, and saw that it had some 270 cost inputs. 'Mine was five,' he told Grigg. 'I needed to have that level of detail.'

Les Tinkler was helping out too. One businessman, Phil Lellman, of the Melbourne-based mine equipment importer Melroad, recalled meeting Nathan and Les in 2003, when they were looking to get their hands on three top-of-the-line Japanese excavators to take back up to the Hunter Valley. 'I remember them walking in,' he told Ben Hills, 'they were like clones – half a ton of meat each, I'll tell you. They looked like a couple of yobbos [with their] big legs and double chins. We had to find a couple of big chairs for them without arms.' But the Tinklers knew their stuff – 'they could talk the talk and walk the walk,' Lellman said – and a deal was done.

It has taken time for it to become clear just how tough these early years were for Tinkler as he tried to establish himself in the mining business. '[W]e had just had our second boy when things got really busy,' he has said, 'and I was travelling a lot but you cope, and thankfully I had a very supportive family.'

At times, his young family was literally broke. In February 2003, Singleton Council took Nathan and Rebecca to court over $1044 in unpaid rates, and in April the Singleton Local Court issued a judgement against them, totalling $1571, including legal costs. The Tinklers couldn't pay, and in May the council's lawyers got writs against them, with the debt rising to $1741.

According to the notice on file, the Sheriff's Office seized just about everything in the Tinklers' home: a Hoover dryer, a wooden buffet/hutch with leadlight doors and three drawers, a sideboard, a three-piece cream check lounge suite, an LG 68-centimetre black TV with a wooden stand, an LG silver video recorder, a Samsung microwave, an LG washing machine, a Kleenmaid dishwasher, an orange Husqvarna mower and catcher, a Stihl whipper-snipper, and a Kelvinator fridge and freezer. The stuff wasn't physically taken away, but Tinkler was ordered not to dispose of it, or face a $5500 fine.

In June Tinkler got a stay of execution, promising to pay off the debt at the rate of $100 a fortnight. He swore a statement on the couple's property and means. It was a picture of ordinariness:

Tinkler was the sole breadwinner, on a wage of $950 a week, plus $50 in child support payments. He and Rebecca owned their home, worth $400,000, with a mortgage of $280,000. The mortgage cost $400 a week, food $200 and bills $120. Tinkler's assets included a car (according to the number plates, it was a blue Subaru four-wheel drive), furniture worth $50,000 and electrical goods worth $20,000. He owned one share, in his own company, worth a dollar. He had $1000 in a Commonwealth Bank account, but owed $3000 to American Express and $8000 to St George Bank – these payments totalled $80 a week.

The Tinklers paid off $550 but by December were falling behind again. In March the sheriffs were back at their door, writing another seizure notice – issued to Rebecca, this time. It may be apocryphal, but long-time Tinkler associates say that when the repo men arrived, Rebecca refused to budge from the couch, shouting: 'You're not taking my fucking couch, or if you are you're taking me with it!' Somehow, according to the file, the family scraped together enough to pay off the debt in April.

By now, though, they had fallen behind on their mortgage, and on 30 March 2004 they were hit with a default notice. In June 2004 Perpetual Trustees got an order in the Supreme Court for the repossession of the Tinklers' home. They weren't kicked out, but in July that year the Singleton house was sold for $420,175. The Tinklers appear to have moved back to Nathan's parents' place at Coolibah Close, Muswellbrook – at least, that was Tinkler's address according to a December 2004 court filing. At the age of 28, and like his father before him, Tinkler had risked everything, and lost. Four years later, when he first topped the Young Rich List, Tinkler admitted:

> I'm not proud of what I've put my family through to get here. I have a young family and I was burning the candle for 100 hours

a week. Life's changed pretty fast, but I guess it just allows you to do the things you want to do. The badge on the front of your car changes, and life goes on …

Tinkler didn't let the foreclosure stop him. Even as he was losing his house, he was rearranging his business affairs. On 19 May 2004 he registered a string of private companies, including the one he would soon rename Custom Mining Ltd and sell for a fortune. There was also Tinkler Investments, with Rebecca as secretary and director, which held his stake in Custom Mining Ltd, and L and Z Tinkler Investments, named after Les and Zelda, who were the company's sole shareholders and directors. From this time on, all Tinkler's major assets would be kept out of the reach of creditors, held either in Rebecca's name or in the name of one of the many private companies he established, often with Rebecca as the sole shareholder, on behalf of their family trust.

The Tinklers moved to Newcastle. In November 2005 Rebecca paid $620,000 for a home on Grayson Avenue in the quiet suburb of Kotara, taking out a new $424,000 mortgage, despite the fact they'd defaulted on their previous mortgage. Curiously, the loan document showed that Rebecca's address was in Charlestown – at a friend's place, as it happened. But for at least the next couple of years the Tinklers would call Grayson Avenue home, and they sent their kids to the nearby school.

Tinkler was a man in a hurry. That same month he lost his driver's licence for six months after he was clocked doing 108 kilometres per hour down Newcastle's main drag, at two o'clock on a Monday afternoon. The incident cracked a small mention in the *Newcastle Herald* – headlined 'Hunter Street Hoon' – Tinkler's first appearance in a newspaper he would come to hate.

*

For much of the 1980s and 1990s, coal prices were stuck in the doldrums, with thermal coal – used for power generation – trading at between US$20 and US$40 a tonne, and generally at the lower end of that range. The then boss of the NSW Coal Association, Meredith Hellicar, coined the term 'profitless prosperity' to describe those years, when Australia sold its coal to Japan too cheaply; the export tonnes kept rising, up eightfold over the two decades, but margins were low and costs had to be contained. The Hunter Valley was not booming. Unemployment hit 12 per cent after BHP closed its steelworks at Newcastle in September 1999, with the gradual loss of 3000 workers.

The *Muswellbrook Chronicle* serves as an interesting barometer of the Hunter Valley coal industry. Each year the *Chronicle*, like the *Singleton Argus*, carries a substantial advertising supplement that promotes 'Coal Discussion Day'; in 1998, the year Tinkler got his trade certificate, it was all about staff cuts, the challenges affecting businesses, and making the best of a bad situation for those workers who'd been laid off.

Production had doubled over the past decade, but employment had only grown by 20 per cent, with jobs peaking at just above 6000. Around 750 jobs had been lost that year at the Hunter Valley No. 1, Mount Thorley and Drayton mines. The demand outlook was flat, and employment was expected to keep falling. 'As Indonesia and China continue to compete against Australia for the export market, and coal prices continue to fall,' the supplement warned, 'there is a significant threat to the demand for coal from local producers.' The Australian Bureau of Agricultural and Resource Economics (ABARE) was estimating that real-world coal prices would fall by more than 2 per cent a year between 1997 and 2010.

How wrong they were. Just three years later, at Coal Discussion Day 2001, the *Chronicle* trumpeted: 'Thirteen new coal projects on the books in the Hunter alone.' ABARE was now sounding more upbeat: after four years of easing prices, in the first quarter of 2001 thermal coal exporters had negotiated a 20 per cent increase with Japanese power utilities, driven by strong demand from new coal-fired power stations there and in Taiwan, Korea and Malaysia. Spot prices had jumped to US$34 a tonne. ABARE speculated that China would not be able to increase its exports as much as it planned, because of its antiquated rail system and the impending closure of unprofitable and unsafe mines.

In fact, the reverse happened: far from increasing its exports, China started to burn more and more of its own coal, gradually withdrawing from the export market. Between 2003 and 2005, coal prices broke out of their two-decades trading band, exceeding US$50 a tonne.

Australians began to realise that they were in the box seat to take advantage of the rapid growth in China and India, which was increasing the demand for key commodities such as coal and iron ore. China was building a new coal-fired power station every week, we heard over and again. 'Stronger for longer' was the mantra, as prices kept climbing and Australia's economic expansion continued; at an earnings result in early 2007, Oxiana chief Owen Hegarty predicted that prices would be 'Stronger forever'!

Tinkler twigged to what was coming. Before he got his foot on a coal deposit, as he told Angus Grigg, he travelled to India and China:

> In India it was more of an appreciation of fuck how far they've got to go because the place is just a shit hole. The only thing that is lower than the conditions are the people. I said OK get me out of here. Everyone I sat with in China was 'OK, we want to buy 50 million tonnes' and I was, 'you mean 50,000 tonnes?' 'No, 50 million tonnes'. I was blown away by the sheer enormity of it.

Impressed, Tinkler picked out some big power users and utilities to speak with in China. He soon decided that everybody back home – including half the coal industry – was missing the big picture.

> The Chinese are active, they want to meet people that can deliver them resources. People would say to me we have 64 power stations and generate more electricity than Australia. I guess I was bullish on where it was all headed. They said we want to buy the equivalent of every [tonne of] coal that is exported out of the eastern seaboard of Australia. I think that scale issue is something that Australia has still not come to terms with. Don't worry about doubling production, you could increase it by ten-fold and not have much of an impact on world markets. Australia is being held to ransom by the BHPs and Rios of the world: it's worth it for them to work harder for less, to keep the prices up. When I went to China I thought, 'The demand is real, this is going to happen.'

It was a long way from Mount Thorley. But not for long. Tinkler's mind was made up: coal was the right game to be in. He was hungry.

2.

MIDDLEMOUNT
FROM $1 MILLION TO $442 MILLION

The loss of the family home was a setback, but Tinkler's eyes were now on a bigger prize. He and Higgins began formulating a serious plan to step up from equipment and labour hire, and use what surplus cash they had to build a new business as investors in the coal industry.

In 2004 Tinkler and Higgins sold their labour hire business to Tesa Mining for about $400,000, plus an 'earn-out'; soon the pair had moved from their Mount Thorley yard to new offices on Newcastle's trendy Boardwalk. But the money they received wasn't enough for them to buy a coal project. If they were going to raise some serious funds, what the two tradies-cum-small businessmen needed above all was cred – someone who could open doors for them at the big end of town.

In September that year, Tinkler and Higgins invited Peter Mallios to dinner at the Sofitel Wentworth, along with Tinkler's father, Les. Mallios had been the acting general manager at Bengalla until 2001, when he'd taken a redundancy and gone to run the western Sydney industrial paint company Vertikote. He'd also been on the interview panel that had appointed Tinkler to his job at Bengalla in 2000. Now, four years later, Tinkler approached Mallios out of the blue to see if he would join Custom Mining as its first chief financial officer, to help evaluate mining projects but also to bring with him his network of potential backers. The tables had turned: Tinkler was interviewing Mallios.

Mallios agreed, accepting a six-figure salary to moonlight for Custom Mining in the expectation that he would also get a 10 per cent cut of any deal that eventuated. For the next year and a half he worked hard for Custom Mining, and somehow also held down his full-time job at Vertikote. He had trouble getting his invoices paid but put that down to disorganisation and Tinkler's usual cash-flow dramas. Mallios also introduced Tinkler to an up-and-coming Ernst & Young partner, Paul Flynn, who would become Custom Mining's external auditor; much later, he too would join Tinkler's team.

Tinkler and Higgins began searching for an asset. They first met with officials from Queensland's mines department in December 2004, but were told that the state's coal resources were 'tightly held with few available opportunities', the government's records noted.

The tiny Custom Mining team sized up a number of projects over the next couple of years. At first they had their eye on the Cameby Downs project, but were beaten to it by another coal junior, Syntech Resources. Custom Mining also looked at Anvill Hill, Jellinbah East, Diamond Creek, Kogan Creek and the Monto project. The most important thing, Tinkler remembered, was 'getting in the car and going up there and having a look around'.

One neglected tenement looked promising. Six kilometres south-west of the town of Middlemount, Mineral Development Licence (MDL) 282 covered almost 2000 hectares of farmland. Anglo American's Capcoal had drilled over a hundred holes there in the 1980s – estimating that the deposit contained some 254 million tonnes of coal – but had relinquished its exploration rights and developed the German Creek and Foxleigh coal mines on either side instead.

The Canadian listed miner Sennen Resources held a majority interest in the site, through an Australian subsidiary, Ribfield, managed by veteran geologist Norman Zillman, one of the original founders of the unlisted explorer Bandanna Coal. Ribfield had applied for MDL 282 in 1998 and had been granted a five-year licence in 2002, but did very little with the site for the next two years, except meet potential development partners. When coal prices started to lift, a Bandanna affiliate, DJB Coal, part-owned by the coal industry stalwart Jeremy Barlow, 'farmed in' to MDL 282 in November 2004 – that is, it agreed to do a bankable feasibility study and apply for a mining lease, in return for a stake in the tenement.

But Barlow thought the drilling results at Middlemount looked 'pretty average, so instead of pursuing the development, DJB decided to sell out immediately, even before gaining a mining lease. An industry broker, Global Resources Asset Exchange, whose director Robert Johansen was on the board of Bandanna, was appointed to sell the project. Seven offers came in by January 2005, with Sennen reporting considerable interest from Indian, Korean, Chinese and Australian companies, but all fell short of expectations.

The Perth-based mining mergers and acquisitions team at Ernst & Young was asked to drum up interest, and a formal memorandum for prospective bidders was released in July 2005, with bids due by September. Again, the results were disappointing. Ernst & Young's memorandum had a touch of 'blue sky' about it, tarting up Capcoal's old 254-million-tonne resource and reserve estimates, which were not JORC-compliant.[1] The fee structure gave

---

1 'JORC-compliant' refers to an ore estimate using the code of practice maintained by the Joint Ore Reserves Committee, an industry body. 'Reserves' are the standard industry measure for minerals or hydrocarbons, and are categorised as either 'proven' or 'probable'. Reserves must be economically recoverable, and have a higher degree of certainty than 'resources', which are classed as 'possible' and categorised as 'measured', 'indicated' or 'inferred'.

Ernst & Young an incentive to seek bids over $50 million – the kind of price Sennen and DJB were hoping for.

The original MDL application in 1998 had anticipated that Middlemount would have a thermal coal deposit, but Ribfield's annual reports to the mines department in 2003 and 2004 said it also contained smaller amounts of coking or metallurgical coal – used for steel-making – and so did the Ernst & Young memorandum. It would take a combination of open-cut and underground mining to get at the coal, however, since three-quarters of the 254 million tonnes lay more than 100 metres deep. This was one of the drawbacks of Middlemount. As Tinkler later told Grigg, 'you had to dig a bit more dirt before you get to the coal and that probably put a few people off'.

Tinkler and Higgins got serious. Though neither was a geologist or a mine engineer, their time in the pits and talking to the guys on the ground counted for something. Tinkler reckoned he knew just how to mine Middlemount, telling people it was just like the Hunter Valley's Mount Owen mine. When it came to coal, Tinkler could talk the talk. He had an amazing ability to absorb detail: one industry executive says that, in those days, Tinkler 'knew every coal asset in Queensland and could name the last three geologists who'd drilled every asset in Queensland'.

From mid-2005 onwards, the partners sought to raise every cent they could – from family, friends and acquaintances, as well as any outside investors they could get in front of – in order to get a hold of Middlemount. Tinkler would later observe that, although Australia is a commodity-based economy, investors are reluctant to put their hands in their pocket for exploration, and 'probably more so if that person only has a Muswellbrook TAFE electrical trade certificate and a bullish attitude'.

Personally, Tinkler was still strapped for cash. He never had money on him, according to those close to him at the time. After business meetings he'd find himself unable to get his car out of the car park. He was living off his credit card and the wolves were never far from the door.

In August 2005, as the first round of the Middlemount sale process was drawing to a close, Newcastle accountants Moylan's Business Solutions initiated a wind-up against Custom Mining for unpaid fees. No sooner was that settled than Tinkler's other company, Custom Mining Solutions, was hit with a wind-up order by Melroad, which was owed $19,000 in sales commission fees for the three excavators it had sold to Nathan and Les in 2003. Melroad now got orders to seize Custom's property, including the computers Tinkler was using to work up his bid for Middlemount. Melroad's boss, Phil Lellman, later told Ben Hills that Tinkler would at that point have 'done anything to stop us taking those computers away'. Melroad got its money, with interest, and the wind-up order was lifted.

The Australian Taxation Office, too, was chasing Tinkler, and in 2005 it successfully sued him personally in the Newcastle District Court. In July he was served with a statutory demand for $214,380 at his Newcastle offices, and in September the ATO also launched wind-up proceedings against Custom Mining Solutions. Tinkler dragged them out as long as possible, then settled – with the help of some clever accounting. It was the first of many wind-up actions the tax office would launch against him. 'The ATO is not to be trifled with,' Tinkler told Grigg in 2010.

Despite the intense financial pressure, the work on Middlemount was getting done. Tinkler and Higgins would sit together in their Boardwalk office, troughing burgers and Cokes as they pored over multi-level spreadsheets on the project. They could not afford to pay expensive consultants and so had to do most of the work themselves, guided by Mallios and John Moore. In August 2005 Custom Mining put in a bid on Middlemount, high enough to get them a period of exclusivity, but they could not complete.

Nor did anyone else, though. In early 2006 Mallios came up with a new plan to finance the Middlemount bid – raise $5 million from fund manager Colonial First State – but Tinkler and Higgins baulked at the amount of equity they would have to give away, which was effectively half the project. They rejected the deal, and by September they had turned against Mallios completely. Two years later he would sue them for breaching an agreement to give him 10 per cent of Custom Mining. Tinkler and Higgins denied the allegations; the dispute was settled on confidential terms.

One of the first to invest with Custom Mining was Richard Jennings, a mine engineer who also worked for a time as the company's operations manager. In June 2006 he had invested $200,000 of his own money, but when Tinkler and Mallios parted ways Jennings also pulled out. He sold back his shares for the same money he'd put in – a decision that would no doubt smart later on. Jennings' sale implicitly valued Custom Mining at $4 million, although you'd have been hard-pressed to find someone who'd pay that much for a company that at the time had no assets.

Custom Mining told the Queensland mines department in May 2006 that Middlemount was its prime target and it was already in discussions with Sennen. In April 2006 Tinkler and Higgins had borrowed as much as they could, mainly from NAB. A string of loans were registered to Custom Mining Solutions, which had owned the labour-hire business and was still Tinkler's cash cow. Two months later, Tinkler sold the old house at Muswellbrook for $318,500, which he tipped straight into the business.

After buying Jennings out, Tinkler owned 71 per cent of Custom Mining, while Higgins owned 24 per cent and O'Reilly's private company held 5 per cent. With everything that Tinkler and Higgins could scrape together – some $1.3 million, Tinkler said later – they now had the semblance of funding.

They couldn't pay their lawyers, however. The Newcastle-based Sparke Helmore had done a lot of work on the Middlemount deal, running up a six-figure legal bill. The legal work had been done partly at the request of some potential investors in Custom Mining, and when they decided not to go ahead, Tinkler simply disowned the bill: it was their problem, not his. It was lawyers at ten paces. Tinkler also ran up a bill to the big Sydney firm Gilbert & Tobin, which launched wind-up proceedings against Custom Mining Solutions. It was a mess which threatened to derail the whole bid. Custom needed top-notch advice but couldn't afford to pay for it.

Tinkler asked around and was told that Philip Christensen, a wily partner at Freehills' Brisbane office, was the gun lawyer he needed, with decades of experience in the resources industry and a blue-chip clientele. Tinkler cold-called him. The much older Christensen found Tinkler refreshing – a young bull having a red hot go, a guy with a big vision but also an ordinary bloke you could sit down and watch the footy with. The two men came to an arrangement: Christensen would go without fees along the way, but would be paid double if the deal completed successfully.

Christensen duly tidied up the legal disputes in Newcastle and set up a dedicated vehicle to bid for the project, Custom Mining (Middlemount), a wholly owned subsidiary of Custom Mining. He also introduced Tinkler to Terry O'Reilly, a former senior executive at Rio's Coal & Allied, who joined the board of Custom Mining as a door-opener, lending a bit of grey hair and respectability to the company. O'Reilly got a 5 per cent stake in Custom Mining for the nominal sum of $10.

The initial exploration at Middlemount had started in August 2005 under a silent purchase agreement between Custom Mining and DJB. Finally, on 16 November 2006, after two years of effort, a deal was done. Although the vendors had doubts about Tinkler's financial capacity, he had been there all the way, as other bidders came and went. Sennen and DJB, which had spent very little on the tenement, had little choice but to give Tinkler a chance. 'He was very persistent,' Jeremy Barlow later recalled. 'He kept working away and finally got us to agree to a deal.'

*

Middlemount was a remarkable buy: the price on the unwanted tenement was cut to a bargain-basement $30 million, split equally between the two vendors, Sennen and DJB, who agreed to accept a non-refundable deposit – an option, really – of just 3 per cent. This amounted to just $1 million up front, or $500,000 each. DJB and Sennen also agreed to an extended settlement, with the purchase price to be paid in three tranches, meaning it would be fully two and a half years before they got all the money owed to them. These generous terms allowed Tinkler and Higgins to bid on an asset they did not have the money for, and then scramble to raise the rest.

After the regulatory hurdles were cleared – a formality in this transaction – the sale contract gave CML eight weeks to drill and analyse the deposit, and another 16 weeks to make the first tranche payment of $9 million. All up, Tinkler and Higgins had about six months to raise a third of the purchase price and complete the acquisition. If they failed, they'd wasted a million bucks.

The partners went into overdrive as they worked to raise the money. Plenty of wealthy private investors looked at going in with them in the Middlemount deal, but baulked. One was Phil Arnall, a senior executive at the engineering firm Bradken, commodore of the Newcastle Cruising Yacht Club and a leader in the city's business community. Arnall had been on the board of Tesa when it had bought Custom Mining's labour hire business. In 2011 Arnall spoke to the *Newcastle Herald*: 'I've told this story, against myself, to all of my friends, how I was the one who didn't make $42 million by not going with Nathan … we made our decision at the time and he went on to prove everyone wrong.'

Tinkler drove down to Sydney to meet various high-profile investors, putting himself up in a hotel. The billionaire Lang Walker reportedly knocked him back, as did the former Peabody managing director Bob Humphris. 'I reckon I knocked on every door in Sydney,' Tinkler told the *AFR* in 2010.

Finally, Tinkler pitched to Barry Dawes at Martin Place Securities, a boutique firm that specialised in raising private money for speculative mining investments. It had backed Wayne McCrae's listed copper hopeful Cudeco, which became an outrageous chart-topper in 2006. Dawes was receptive, and ultimately joined the board of Custom Mining. 'I think they were impressed I'd put the $1 million down,' Tinkler recalled. 'They thought this guy is either a complete idiot or he is going to come up trumps.'

Martin Place and some two dozen clients invested $3.7 million in two tranches between October and January 2007 – first at $5 a share, then at $10 a share – to take 26 per cent of CML. As each new investor paid more per share than the last, the implied value of Custom rose accordingly.

With 2.7 million shares on issue, and shares selling for $10 each, the unlisted company was theoretically worth $27 million. If their own shares could have been sold at that price, Tinkler's 52 per cent stake would have been worth $14 million, and Higgins' 18 per cent stake $5 million. But it was paper wealth: shares in an unlisted company with nothing to its name but an option over an undeveloped coal project.

All the money from Martin Place went into proving up the deposit. Higgins coordinated the drilling campaign, his steadiness complementing Tinkler's drive. Tinkler, who'd stayed in the Middlemount pub, recalled taking core samples from Middlemount to Gladstone and back, jumping in his Hyundai Santa Fe for the 600-kilometre-plus round trip. He kept what he was doing pretty quiet. 'You can imagine that once I paid that million dollars, everyone was like, "That's a piece of shit," running it down in the market place ... I told the driller and geos that "We are just working there".' Nevertheless, Tinkler told Grigg, this was 'one of the great times of my life. I was under enormous pressure but [had] enormous satisfaction.'

The results came up trumps: Middlemount had coking coal, and of much better quality than either Sennen or DJB had realised. Tinkler and Higgins were on a winner.

Custom Mining started to expand. Short of expertise, Tinkler hired mining engineer Peter Bannister and Tom Todd, a bright young English accountant with a first-class physics degree from London's Imperial College, who was working in the resources division of the lender GE Capital. Tinkler had come across Todd when he'd approached GE for funding. GE hadn't stumped up, but he and Todd had hit it off. Todd was a plucky guy who liked Tinkler's sense of humour, his willingness to hurl convention out the window, and his guts. They shared a love of fast cars. Todd went further than Christensen, chucking in his job to join Tinkler in April 2007, right after the birth of his second child.

Meanwhile, Tinkler and O'Reilly went on an international roadshow, targeting potential Asian investors. They got plenty of knockbacks but one company took the bait: the Hong Kong-based Noble Group, one of the top-10 commodities traders in the world.

Noble, which was listed on the Singapore Stock Exchange, was building a significant stake in Australia's coal industry and had an office in Newcastle. Noble was keen to support junior miners who could provide an alternative source of commodity supply to the big diversified miners such as BHP and Rio. Back in 2004, word about the upstart Tinkler had reached William Randall, the head of Noble's energy division. Randall himself came from

Muswellbrook. His father, Colin, was a coal industry veteran, a mine engineer who had edited the weekly *Hunter Valley Coal Report* for decades and was a director of a string of junior coal companies.

An intense, balding workaholic who had studied marketing and finance, Randall had stepped up to run Noble Energy after his boss died suddenly of a heart attack. Based in Singapore, Randall constantly travelled the world's coal provinces, from Australia to Indonesia to Mongolia to Russia. Randall kept his finger on the pulse and had a laser-like focus. 'My life is simple,' he wrote in one email to Tinkler, 'my family and Noble.' Indeed, he had never worked anywhere else. Tinkler's relationship with Randall would track the highs and lows of his career over the next six years. Tinkler recalled that once the drilling showed Middlemount was coking coal, Randall hadn't hesitated, saying, 'Right, we're in.'

To get the Middlemount deal over the line, Tinkler, O'Reilly and Todd flew to Hong Kong to meet Noble's chairman and largest shareholder, the multibillionaire Richard Elman. He was a self-made man who had left school at the age of 15 to work at a scrap-metal yard in London, and had been trading hard commodities ever since. Elman was wary of Tinkler; he wanted to see the whites of his eyes. But Tinkler was persuasive, O'Reilly's presence was reassuring, and Randall vouched for the Australians. Elman signed off on the deal.

On 4 April 2007 Noble agreed to pay $25 million for 25,000 shares of Custom Mining (Middlemount) – roughly 10 per cent of the project – and another $65 million for 71,450 new shares, which eventually would take its total stake to 30 per cent. Both commitments were staged – only $10 million and $49 million were to be paid up-front, with the rest to come in tranches over the next year. Noble also agreed to extend CMM an $80-million loan to fund the development.

The deal was a massive coup for Tinkler. Having effectively paid $90 million for 30 per cent of Middlemount, Noble was valuing the project highly, at $300 million. Six months earlier, Noble could have had the whole project for a tenth of that. Custom's share of Middlemount was now worth $210 million.

Between November 2006 and April 2007, the value of Tinkler's own stake in Custom Mining had jumped from under $750,000 to $110 million. More than any other transaction, it was this deal that made Tinkler rich. Noble was a sophisticated investor in the coal industry, and although shares in Custom Mining were highly illiquid, having Noble on board gave the company credibility in the market. Tinkler and Higgins and their small group of backers had made the grade. And all this before the purchase had even settled, and the first tranche payment been made to Sennen and DJB.

Noble took its pound of flesh, getting marketing rights to the coal from Middlemount and a 4 per cent royalty on every tonne sold once the mine was up and running. (In a trademark piece of legal advice from Christensen, Tinkler and Higgins kept a royalty worth $1 for every tonne of coal eventually sold from Middlemount, just in case everything else turned south, as it later did.) Noble also had an option to increase its stake in Middlemount by another 20 per cent, for another $100 million. Noble took security over the whole tenement, as well as two seats on the board of Custom Mining (Middlemount). Noble inserted an important rider into the agreement: neither Tinkler nor Higgins was to sell down their combined stake in Middlemount below 50 per cent until six months *after* construction was finished.

At the stroke of a pen, the Noble deal turned the Middlemount mine from dream to reality. Not only would it allow Custom Mining to make the first $9-million tranche payment to Middlemount's vendors, DJB and Sennen, and complete the acquisition, but it also meant they could buy out the cattle grazier who owned the land. Custom put up bonds of $10 million for the mine's builder,

Leighton Contractors, which was soon awarded the $500-million contract to start digging the open-cut. Another $30 million was pledged to the engineering firm Sedgman, which would build the coal wash plant for $65 million.

The Noble deal also allowed Custom Mining to do a selective buyback of shares, diluting the Martin Place investors, who, Tinkler and Higgins felt, were doing a bit too well out of the deal. Martin Place investors were offered $25 a share; roughly half its investors sold. They therefore made five times their money in about six months – not a bad return. Those who hung on – no doubt including some who could do the maths and work out the value per share that was implied by the Noble deal – would soon do much, much better. All in all, Custom Mining bought back 336,000 shares, or 12 per cent of its total issued capital, for $8.4 million; now the Martin Place investors owned only 16 per cent of the company. Tinkler's stake jumped back up to 60 per cent, while Higgins' and O'Reilly's shares increased to 20 per cent and 4 per cent, respectively. Tinkler complained that it cost him 'a fucking shitload to get [the Martin Place investors] out. While we got a killing they certainly did also.'

After the Noble deal was agreed, but before the funds had begun flowing and the Middlemount purchase was completed, Custom Mining Solutions was hit with another wind-up action, this time by the insurance company GIO, over unpaid workers' compensation insurance premiums. Tinkler later told the *AFR*: 'The sheriff was at the door. I had dug myself in so far, but failing was not an option.'

The whole Middlemount deal was in the balance and could have gone either way. At the last minute, the day before the Noble agreement was signed, Custom was able to scrape together the funds to settle GIO's claim. At last, the millions of dollars from Noble arrived and the purchase of Middlemount could proceed.

The payments were not made to Sennen and DJB until eight p.m. on settlement day, Thursday 21 June 2007 – a red-letter day for Tinkler, Higgins and their small crew. They headed out to celebrate, but by the time all the loose ends had been tied up it was so late that the only thing open in Brisbane was the McDonald's at the bottom of Riparian Plaza. High-fives and burgers suited Tinkler and Higgins just fine. The rest of the world had no idea what an outrageous amount of money they'd just made.

*

Queensland premier Peter Beattie was all set to announce the new mine, then his minders checked with the department. The advice that came back was sceptical: Custom Mining did have significant experience in New South Wales and had done $5 million worth of exploratory work at Middlemount, but had not yet applied for a mining lease and had not yet taken control of Ribfield. The approval was likely to take longer than they expected. There would be no announcement just yet.

But who cared? Middlemount was just the start. The Noble deal freed up Custom Mining to fund some of the other projects it had been investigating. Indeed, the very next day, 22 June, Custom paid $2 million to Macarthur Coal for an option on Monto, another Bowen Basin tenement.

At Monto, as at Middlemount, two joint-venture partners were at odds. Macarthur Coal and the powerful QCoal Group owned the project 51–49, but Macarthur was dragging its feet on development. Tinkler was hoping to pick Monto up cheap, sort out the mess with QCoal and flip it for a motza.

The Monto negotiations brought Tinkler into the orbit of Ken Talbot, the Macarthur Coal founder who would cement Tinkler's fortune and whom Tinkler would come to consider as a mentor. Talbot, then in his mid-50s, was a truck driver's son who had started out as an engineering cadet for BHP in the Illawarra. After a few years at the helm of Bond Coal in the heady 1980s, Talbot had made his first millions as part of a consortium that developed the Jellinbah East coal mine in central Queensland, before busting up acrimoniously.

Talbot was a great risk-taker. Having founded Macarthur Coal, he had built the Coppabella and Moorvale mines, targeting the Bowen Basin's Rangal coal measures, rather than the better-quality Moranbah seams, which were heavily exploited and getting deeper and more expensive to mine. Macarthur carved out a new niche in the coal market, marketing the so-called pulverised coal injection (or PCI) coal, which had low volatility and could be used in the latest Asian blast furnaces to make steel, as a supplement to the pricier hard coking coals.

Talbot's success in this gave him a healthy disrespect for the big incumbent operators; Coppabella itself had been passed over by BHP in the late 1980s. He preferred to stay small, using contractors wherever possible; his business model called for a low capital spend and a low footprint. When Macarthur floated in 2001, with Talbot as its largest shareholder, he became one of the richest men in Queensland. He used part of that wealth to sponsor the Brisbane Broncos and their 'supercoach', Wayne Bennett. In the good times that followed, Tinkler watched quite a few games from Talbot's box at Suncorp Stadium.

In November 2006, however, Talbot was forced to stand down as head of the company he had founded, amid an inquiry by the state's Corruption and Misconduct Commission into $360,000 worth of payments he'd made over a three-year period to then Health and Primary Industries minister, the ALP's Gordon Nuttall, who was later found guilty of receiving secret commissions and jailed. Talbot considered the payments a loan, help to a friend in need. Nuttall's unconvincing explanation was that was he trying to prepare for a life after politics. Talbot, who himself was charged by the CMC in January 2007, retired permanently in June. His deputy chief executive and former chief finance officer, the 37-year-old Nicole Hollows, took over the reins. Talbot remained Macarthur's largest shareholder, owning 36 per cent of the company, and a director, but from that point on he had one eye on the exit. In July he sold an 8 per cent stake to Macarthur's Chinese joint-venture partner CITIC for $113 million.

In mid-2007 coal prices took off, with both thermal and coking coal hitting three-year highs above US$70 a tonne and US$175 a tonne, respectively. They kept going up. A huge adjustment loomed over world coal markets: China, the world's biggest coal producer, was turning from being a net exporter to a net importer of coal. Australia exported 3 million tonnes of coal to the Chinese mainland in 2007–08, and 18 million tonnes in 2008–09. It's possible that Noble, desperate to put its foot on the supply of one of its key commodities by ramping up its activity in Australia, had seen the wave coming.

Maybe Tinkler did too. Work was now starting on Middlemount – a decent-size box cut was in place even before a mining lease had been granted – but Tinkler and Higgins sensed that there could be a quicker way to 'monetise' the project.

From mid to late 2007 there was increasing friction between the two men, possibly over the use of company funds. Custom Mining's accounts showed $4.6 million in loans to shareholders as at 30 June. While it is not unusual for principals whose wealth is tied up in a private business to borrow from their own companies, this was a little strange, given that Custom had been scrounging for every dollar just ten days earlier. The accounts do not show which shareholders were having a lend, but it is extremely unlikely that Terry O'Reilly or Barry Dawes, as independent non-executive directors, sought loans from the company, and inconceivable that any of the Martin Place investors would have done so.

At this time, Tinkler, Higgins, Todd and Bannister were working out of a rented Servcorp office in Brisbane's Riparian Plaza. Tinkler lived in a rented flat at inner-city New Farm, and would share it with Todd, who was commuting from Sydney. Tinkler became increasingly suspicious of Higgins. A division was emerging: Tinkler and Todd were doing the deals, while Higgins and Bannister were looking after the operations. Tinkler also began to feel he was carrying O'Reilly, who resigned in September. Greed was kicking in, splitting the partners.

As Tinkler recalled it, Higgins and O'Reilly came to him saying they wanted to get out. 'I said, "Fuck, what are you looking at me for … I don't have any cash, you know how fucking broke I am."' Then, said Tinkler, '[Higgins and O'Reilly] started hocking their shares around the market and I thought, "Fuck, I am going to be left with an Indian partner … somebody who has got money and the tail is going to wag to dog." So I decided the only thing I could do was sell … we were arguing about price.'

Anglo American, which knew the tenement from its earlier exploration, did due diligence on Middlemount in August 2007. Tinkler went to see Macarthur Coal, meeting Ken Talbot for the first time. Macarthur was a natural buyer. Like Macarthur, Middlemount was targeting the Rangal coal measures and had broadly the same mix of coal, and Macarthur also had excess port capacity. The catch was that Tinkler and Higgins had agreed with Noble not to sell down their stake in Custom Mining until at least six months after the wash plant had been built – and the construction work hadn't even begun yet.

Not being sticklers for the rules, Tinkler and Talbot did a verbal deal anyway. When Tinkler rang Noble to break the news, Will Randall was incredulous.

'You can't do that,' he said. 'Macarthur has a board, and shareholders.'

'Ken moves fast,' Tinkler replied. 'Call Nicky Hollows.'

He wasn't lying. Commercial negotiations had begun in earnest from early November, with Macarthur sending in a due diligence team. Tinkler was acutely aware that the rising coal prices – actual thermal coal sales were now being reported above US$90 a tonne – had increased the value of Middlemount since June. In fact, the rising market had probably added more value to the project than Custom's development work; although construction had started, a mining lease had still not been granted.

Noble had the right to veto any deal between Custom Mining and Macarthur Coal, but the proposed sale put the investor in a quandary: who to get in bed with? Instead of being in a joint venture with a junior private developer, which was obliged to and dependent on Noble for funding, it would be in partnership with a substantial, publicly listed miner whose priorities might be completely at odds with its own. On the other hand, Noble had to question whether Tinkler and Higgins, having moved so quickly to sell out of their joint venture, were reliable partners.

Tinkler's priorities were apparent: he was already spending up big. In September that year, Rebecca paid $5.2 million for a mansion on a 10-acre block at 421 Grandview Road, Pullenvale, on Brisbane's western fringe. A glitzy mass of downlights and beige, the sprawling 1600-square-metre residence had everything: seven bedrooms, an 18-seat cinema, a 3500-bottle cellar, separate quarters for guests, a nine-car garage, an office, a gym, a teppanyaki grill and even a beauty salon. Outside was a tennis court, stables, a golf green, and a pool – complete with waterslides – and an outdoor bar with beer on tap. It was all showy, rather than chic.

Noble believed it had grounds for action against Tinkler and Custom Mining for breaching their April agreement, which might in turn give it the opportunity to exercise its security over Middlemount, snatching the tenement for a song. But that would involve lengthy litigation.

Instead, Noble consented to Macarthur's acquisition of Custom Mining and struck a side deal over Monto, which had descended into litigation in October with the filing of a $69-million claim by QCoal and its partners. By then, Custom Mining's $2-million option was about to lapse – Tinkler had done nothing with it.

In the lead-up to the Custom Mining acquisition, Macarthur's shares jumped, which improved its ability to pay with its own shares. In late November, Macarthur signed a confidentiality agreement with Tinkler, and on 10 December the company went on a trading halt and made an announcement to the stock exchange: Macarthur would pay $265 million for 100 per cent of Custom Mining Ltd, taking over Custom's 70 per cent interest in Middlemount, as well as its next-to-worthless interest in another tenement, Dingo West, acquired in late November.

Most of the Custom Mining purchase was done with Macarthur scrip: $200 million worth of shares, which worked out to 25 million shares at $8.07 each. It was very cheap stock: Macarthur shares had risen almost 40 per cent from $7.43 in late November to $10.27 just prior to the announcement.

Tinkler saw what was happening and decided to take as many shares as possible. Knowing that both Higgins and O'Reilly wanted out, he moved fast. Tinkler got a short-term loan and, a day before the deal was to be announced, offered to buy the two men's incoming Macarthur shares – by some accounts for $9 each, a premium to the share price ascribed under the deal – so they could take straight cash out of the transaction.

It was never disclosed exactly how much money Higgins and O'Reilly took away from the sale of Custom Mining, but it is possible to work backwards and come to a rough estimate. Had Higgins sold his Macarthur shares to Tinkler at $9 each, he would have made $45 million, and on top of that his share of the cash was about $13 million, making a total payout of $58 million from an initial investment of about $250,000. It was enough for anyone. At 38, Higgins was happy to take the money and pull up stumps.

Where some people aspire to be rich and famous, Matthew Higgins aspired simply to be rich and anonymous. Almost from this day on, he would slip back under the radar – until a dispute with Tinkler years later put him, uncomfortably, back in the spotlight. In the meantime, he had no public profile. He did no media interviews, and there were no photos of him on the party circuit. He all but disappeared.

Terry O'Reilly, who owned 4.8 per cent of Custom Mining at the time of the sale to Macarthur, took cash of about $12.5 million – a handsome payout for less than two years' work, which set him up for a comfortable semi-retirement after a long career in mining. A year later, he joined the board of Macarthur Coal as an independent director.

For Barry Dawes and those Martin Place investors who hung on, investing in Custom Mining proved to be one of the best deals of their lives. One experienced investor who had paid $300,000 for 50,000 shares got back $5.8 million before fees. This represented 17 times his money in the space of a year – his best per annum return in more than three decades of investing. This man had refused to sell at $25 a share, on the principle that one should never sell to the owner of a business, who is bound to know something you don't. Tinkler would later complain to the *AFR* that he 'never so much as got a phone call off Barry, to say what a super job'.

By paying $265 million for 70 per cent of Middlemount, Macarthur had attributed a value of almost $380 million to the project – higher again than the $300 million Noble had ascribed some eight months beforehand. Unlike Noble, Macarthur was a miner, and a year earlier they could have had the whole project for $30 million. But by December 2007 the coal industry was in a frenzy: that same month, Anglo paid $712 million for a majority stake in the nearby (operating) Foxleigh mine – a price that made the Middlemount deal look fair.

Macarthur CEO Nicole Hollows would later acknowledge that her cashed-up company was at this time a bit like a kid raiding the lolly jar – it bought too much, and in some cases overpaid. But the Middlemount valuation was supported three years later, in 2011, when Peabody paid $4.8 billion for Macarthur Coal. Of that purchase price, roughly a billion dollars was attributable to Macarthur's half-stake in Middlemount, which by then was up and running after some $400 million had been spent on construction. Whether Peabody would ever recover that billion dollars is a moot point, but for Macarthur Coal's shareholders Middlemount proved a pretty good deal. Noble, too, did well out of Middlemount, although its return was less spectacular than Tinkler's.

The funniest thing? When Custom Mining was sold to Macarthur, poor old Sennen and DJB were still waiting for the remaining two-thirds of the Middlemount purchase price. It would be two years before they would get their next $18 million, while the Custom Mining boys were already in and out for ten times that amount. Tinkler and his team laughed about that for ages.

In one respect, Tinkler did get really lucky: only days after the sale of Custom Mining was completed on 15 January 2008, central Queensland was hit by massive floods. Middlemount was quickly underwater – the pit was flooded, and Leighton's trucks and excavators were sunk – and the project was delayed for years. Had it still owned the project when the floods hit, it is doubtful whether Custom Mining would have been able to bear the additional costs.

Macarthur declared force majeure on 25 January, and suspended its coal supplies. Coal prices spiked again and would soon reach all-time highs, with thermal coal peaking above US$125 a tonne and hard coking coal topping US$300 a tonne – prices way above anything seen before or since.

*

Nathan Tinkler wasn't done making money yet. As was disclosed to the ASX in February, Tinkler Investments (sole shareholder: Rebecca Tinkler) was now the second-largest shareholder in Macarthur Coal, with 22 million shares, or a 10 per cent stake that was worth $178 million at the deal price. In early 2008, the value of these shares was rising fast. Tinkler could borrow freely against his investment in this Top 200 listed company, and did. He was liquid, and he was about to make a splash.

With such a stake in Macarthur, Tinkler had a good claim to a board seat. In the early days after the sale, he saw boundless possibilities – right up to taking control of the company – and hoped to join the board, as Ken Talbot was about to step down. Tinkler wanted to ramp up Macarthur's production to 20 million tonnes a year, and got a scoping paper together. Talbot was behind him, Tinkler claimed. '[Macarthur] had a strong balance sheet, no debt,' he said. 'I thought it was time to put the whole fucking thing to work and Ken was very supportive of that.' Talbot invited Tinkler along to meet the board in early 2008, while he was still a director. It did not go well.

As Tinkler later told Grigg at the *AFR*: 'I really wanted to light a fire under the place, but at the first board meeting they basically told me they were putting the [Middlemount] project on hold. There is about as much mining capacity on that board as you would find in any pre-school.'

Macarthur Coal's chairman, Keith De Lacy, a close friend of Talbot, did not take too kindly to Tinkler either. De Lacy was a former state treasurer, under Labor premier Wayne Goss, and had gone into business after politics in 1998, taking board seats on high-profile Queensland companies including Queensland Sugar and the Cubbie Group. He'd been chair of Macarthur since it floated in 2001. In his late 60s, he was not about to fall at the feet of a 32-year-old who suddenly reckoned he knew everything about mining.

As De Lacy told the ABC's *Four Corners* in 2013, a board seat for Tinkler had definitely been on the cards, but the matter never progressed to an invitation. 'I think he was going into that, if you like to call it, "the arrogance phase",' De Lacy said. 'He was reported as saying things about us, that our staff were no good, and yet for the next three years he was poaching our staff as though they were the greatest staff in the world. So that was Nathan … he wasn't impressed with us as a board. There's probably an element that we weren't impressed with him.'

De Lacy believes Tinkler couldn't deal with his newfound wealth. 'He got too rich, too young. Now, I know that's a problem a lot of people would like to have to deal with, but … it's not that easy. You have to have the psychological adjustment and I don't think he had. So, it went to his head a bit … maybe a big bit.'

Tinkler didn't muck around, deciding immediately that he was going to sell his stake:

> It was all taking shape, then the board got absolutely nervous that things were going to change and it was not going to be the pensioners society that it [was], and so they panicked and pulled the pin. When they said they were not going to go ahead, I said, 'Well, right, I'm out and my shares are for sale.' I said, 'The idea of having my family's wealth tied up with … you lot just fucking gnaws at me, and so I am just going to have to do something about that.'

The decision was made easier by the soaring Macarthur share price, which, after a minor slump around the time of the floods, had taken off again amid rampant speculation that somebody – Xstrata, Anglo, the Indians, the Chinese – was making a bid for it. Macarthur went back over $10, then $15.

Tinkler told Talbot he was a seller, and Talbot decided to help him out – and to help himself by doing so. The pair decided they wanted $20 a share for their holdings in Macarthur Coal. Tinkler's stake became the bait for anybody interested in a full takeover.

Tinkler tried to engineer a takeover bid from Xstrata, contacting local chairman Peter Coates. He threw out the possibility of a sale at $15 or $16 a share, and said that Talbot, too, was a seller. Xstrata said it would only be interested in buying a strategic stake if it was the prelude to a full takeover.

A tight period of four weeks' due diligence was arranged, and Xstrata's bid team was shown Macarthur's books, given presentations and taken on site visits. Hoping to ramp up the share price, Tinkler leaked to the financial press that Xstrata was doing due diligence on a Macarthur bid, but the company's bean-counters couldn't see any value past $10 a share. Xstrata played a waiting game, hoping some of the sizzle would come out of the Macarthur price.

Tinkler told Xstrata there was another potential buyer in the background. They thought he was bluffing but he wasn't. Xstrata didn't know it, but Ken Talbot had asked his right-hand man, Dennis Wood, who handled his private finances, to go out and get offers for his and Tinkler's combined stake at $20 a share. Remarkably, after a quick tour of Asia, Wood had done just that, attracting genuine interest from Korea's POSCO and ArcelorMittal in India.

On 21 April Macarthur announced to the stock exchange that it had received an approach from an unidentified third party 'in relation to a potential transaction'. Macarthur's shares leapt again, hitting $16, then $17. Amid the excitement, Tinkler told the *Courier-Mail*, 'the same people who have approached Ken have approached me and whatever we do, we'll do it together. The ball is in Ken's court and I'm being taken along for the ride.' All Tinkler had to do was hang on. Two weeks later, Macarthur confirmed that discussions with the same party were ongoing.

On the evening of Monday 19 May 2008, Tinkler took the luckiest phone call of his life. By his own account, he was sitting in his Brisbane home, watching television. It was Macquarie Bank, telling him an Indian company wanted to buy his stake in Macarthur for $20 a share and had just transferred $650 million into a trust account.

> I said, 'Is this a fairytale?' I said, 'Fucking Indians, mate, I have done a shitload of time in India, I know all the cunts, and the chances of getting money out of the pricks is fuck all and none.' They wanted me to fly to Sydney in the morning and sign the forms. I said, 'I'm not going anywhere, you can shove it.' I got into the office the next morning and there is a fax – fuck, who sends a fax? – anyway, there's a fax showing $650 million in a trust account, and if I wanted $450 million of it I had to sign the share form and send it back. By quarter past nine my shares were gone and I had $445 million bucks ... unreal!

The buyer, Macarthur revealed the next day, was ArcelorMittal, the world's largest steelmaker, which had taken 15 per cent of the company: Tinkler's entire holding of 10 per cent and another 5 per cent owned by Talbot.

Two days later, a one-page announcement to the stock exchange declared it to the world: Nathan Tinkler had hit the jackpot. In carefully handwritten capitals, right down to the cent, Tinkler issued his 'notice of ceasing to be a substantial shareholder'. Tinkler

Investments, based in Newcastle, had sold 22,094,851 shares in Macarthur Coal for $441,897,020 cash.

Tinkler himself admitted to Grigg that 'for probably 12 months after that I really struggled with the enormity of having that much cash, and it come with a lot of pressure, like media. I was a nobody, then all of a sudden everybody was interested in how you live your life and what you do.'

Tinkler's sale threw Dennis Wood into a frenzy: Nathan was out, but Ken Talbot was still in, left with a 22 per cent holding. Macarthur now announced that Mittal was *not* the third party that it had been in negotiations with since April. The battle for Macarthur was still on.

Within six weeks, Talbot would offload another 5 per cent to Mittal, and a 10 per cent stake to POSCO, all at $20 a share, leaving him with just 5 per cent of Macarthur. Wood played a masterful game, setting the Indian, Korean and Chinese suitors off against each other, and leaving none with a decisive advantage. Wood did such a good job of selling Talbot's stake, garnering $636 million, that he got a 5 per cent bonus, itself worth tens of millions of dollars.

Their timing *had* been exquisite. Macarthur Coal peaked at $23 a share before dropping like a stone, as the flood-induced spike in coal prices came off, the takeover battle wound up in a stalemate, and Macarthur was forced to lower its profit expectations due to supply interruptions. By the end of 2008, in the eye of the maelstrom of the global financial crisis, Macarthur was trading at below $3 a share.

Tinkler believed the success was all his, but he'd had plenty of support along the way. In a pattern that would recur again and again, he had burned many of those he used on the way up. He'd already fallen out with John Moore, Peter Mallios, Richard Jennings, Matthew Higgins and Terry O'Reilly – they'd all had enough. Even

Philip Christensen, who'd taken a big risk to work with Tinkler, and despite the pivotal role he'd played in getting the Middlemount deal over the line, had to threaten to sue Tinkler to get the doubled-up fees he'd been promised. Freehills had drawn up its claim and was about to exercise its solicitors' lien over Tinkler's assets when Tinkler finally paid up.

But who cared? Tinkler was made. It had been almost four years' work, but, against all the odds, he had pulled off one of the most extraordinary business deals Australia had ever seen. Tinkler cracked *BRW*'s Rich 200 list in May 2008, debuting with a fortune of $426 million, and was crowned the richest Australian under 40 in the Young Rich list in September that year. He was now sitting on a fortune of over $442 million. The fun was just beginning.

3.

ALL TOO HARD?
THE STORY OF PATINACK FARM

With his newfound wealth, Nathan Tinkler could now afford to indulge one of his true passions, his first love, horseracing. He was a born punter; it was in his blood. His grandfather Norm had trained horses near Grafton, and his father and uncle were both keen punters and took Tinkler to the races from an early age.

In one interview Tinkler recalled how, as a kid, he always had a form guide in his back pocket; he'd skip classes to duck down the TAB if he 'had one running'. He dreamed of achieving what the Ingham family had done over 40 years with their Woodlands Stud, as he told the *AFR*'s Angus Grigg in 2010: 'the thing that stoked me up ... was looking at those Woodlands guys as a kid and thinking they must be having the time of their life, having a runner in every race in Sydney and half a dozen in Melbourne.'

Tinkler was buying horses almost as soon as he left school. His former supervisors at the Bengalla mine remember him setting up a syndicate with a few of the workers there. In 2002 Nathan's father, Les, had started his own stud, Serene Lodge, at Kendall, near Port Macquarie, where he planned to train horses and eventually breed them.

In December 2004 both Nathan and Les were taken to court by Coolmore Stud – another prestigious farm in the Hunter Valley – for failing to pay a $5500 service fee for one of their stallions, Mull of Kintyre, which had covered the Tinklers' unremarkable mare

Subtle Miss. Neither Nathan nor Les turned up to court, and a default judgement was recorded against them in April 2005.

It was one thing to have big dreams in racing, but it was another thing to suddenly have the means to achieve them. After pulling off the Middlemount deal, and with Les always watching over his shoulder, Tinkler went on a horse-buying spree unlike anything seen in Australia before. No one had ever bought so many horses, so quickly. Tinkler's ultimate ambition was to race the progeny of his own farm.

Through 2007, Tinkler and Les had been quietly buying horses under the Serene Lodge banner – predominantly weanlings, or foals. Sales records tally Serene's buys that year at 71 horses, costing almost $2 million combined. At an average of $27,000 per horse, the purchases slipped under the industry's radar.

In December that year, reports trickled back that an Australian agent, Roger Langley, was spending up at the Arqana mixed bloodstock auctions in the plush seaside resort of Deauville, on the Norman coast of France, the country's main horse-breeding region. Known as the 'Parisian Riviera', Deauville is possibly the most glamorous place in the world to found an instant racing dynasty.

Langley bought fillies and broodmares for a new bloodstock venture – mistakenly reported as 'Potinack Farm near Brisbane' – including the three-year-old mare Jambu, which was bought for $1.2 million. If Tinkler was there, nobody noticed. Jambu was from the family of Coolmore's star sire, Encosta de Lago. Langley told the UK's *Racing Post* that her value rested entirely on her breeding, not her racing potential: 'I love that family because it can upgrade a whole operation,' he said. Media reports put Tinkler's spend at the Arqana sales at $2.6 million for five horses – three fillies and two broodmares.

Now Tinkler needed somewhere for his horses; Serene Lodge was never going to be good enough. Just before Christmas,

Andrew Bowcock, the owner of the prime Alanbridge Stud – near the town of Scone, right across the Hunter River from Woodlands – got a call from an agent wanting to put an offer on his property. 'It's not for sale,' said Bowcock, a third-generation studmaster whose family had been in the area since the 1940s. The answer came back: 'Everything's for sale at the right price.'

The buyer was Nathan Tinkler, and he would not take no for an answer. He offered $8 million; Bowcock later described Alanbridge as 'the dearest bit of dirt ever sold in the Segenhoe Valley'. Bowcock recalled showing Tinkler around the property; 'he never even walked into the house before he bought it,' he noted. But Tinkler didn't care about the house – he wasn't going to live in it.

To fit all his horses, Tinkler soon added the nearby Riverslea Stud, bought for another $3 million, and combined them as Patinack Farm. It was named after his grandfather's stud in Ireland, Tinkler told reporters – although it seems unlikely that this was Norm, who was born at Webber's Creek, near Maitland; it was certainly not mentioned in his obituary.

Tinkler managed to keep these property purchases quiet for a while, but his next move grabbed headlines on both sides of the Tasman. On 28 and 29 January 2008, at the Karaka Yearling Sales in New Zealand, the hitherto unknown Tinkler stunned the crowd, with the hitherto unknown Patinack Farm buying horse after horse, at big prices. Patinack was the top buyer at the sales, paying $7.9 million for 24 horses, including the top price: $925,000 for One Cool Cat.

Quoted in New Zealand's *National Business Review*, Tinkler sounded chuffed: 'This filly took everyone's eye and we had to beat off a bit of competition to get her. At the end of the day she is a half-sister to Royal Asscher and if you are going to get a filly out of that family, well, it was always going to cost us a lot of money.'

He also outlined the Patinack strategy: 'Business has been kind to me, and now I have a bit more time to play with horses. We're trying to build a broodmare band and looking at potential stallions, and we're aiming to spread the risk a bit by buying them as yearlings.'

Holding Tinkler's hand at Karaka was Anthony Cummings, son of the Melbourne Cup legend Bart Cummings and a seasoned Sydney trainer himself. Tinkler had engaged him as an adviser; they would quickly buy more than a hundred horses together, including Patinack's foundation stallion, Casino Prince, which was bought from Cummings himself. It was a private deal, but whatever Tinkler paid, Casino Prince was a good buy: he would soon become a Group One race winner and then sire All Too Hard, a half-brother to Black Caviar. More big-name stallion buys followed, including Wonderful World – a multimillion-dollar deal – Teranaba, Murtajil and the Argentinian colt Husson.

Next, Tinkler needed stables. He took out 90 boxes at Randwick and 35 at Caulfield. He spent $3 million to buy a 60-stable complex at Bundall from the Gold Coast Turf Club's chairman, Hoss Heinrich, and soon upgraded it.

On a roll, in March 2008 Tinkler and Cummings turned up at Gerry Harvey's Magic Millions auctions on the Gold Coast and splurged $19 million on 59 horses, including record prices of $2.2 million each on Metallurgical and an unnamed colt by champion stallion Redoute's Choice out of Gypsy Dancer (it was later named Foxtrot Oscar).

By now, everyone in racing was taking notice. The *Gold Coast Bulletin* ran a lengthy profile of the mystery 31-year-old who by now had splurged $38 million in three months. Much more than the sale of Custom Mining to Macarthur Coal, the racehorse splurge marked the beginning of Tinkler's notoriety.

Les Tinkler, on the phone from Port Macquarie, was quoted at some length in the *Bulletin* piece. He noted that the curiosity about Nathan was 'understandable with the amount of money he's spending'. His own role was less clear, and he seemed to distance himself:

'I am a part of what he does ... but I am older than he is and I never planned to get in as deep as he has. But that is his choice. He always wanted to go a long way into the [racing] industry and he is certainly doing that now.'

In quick succession, Tinkler and his rapidly growing entourage bought another four fillies at the Perth Magic Millions, for $148,000, then 24 yearlings at the Melbourne Premier for $3.5 million, then (through Cummings) 18 yearlings for a reported $7.6 million at Inglis's Easter Sales in Sydney. Ahead of those sales, Tinkler told the *Sydney Morning Herald* that not even the presence of Dubai's Sheikh Mohammed bin Rashid Al Maktoum at Inglis's Newmarket complex concerned him: 'we all go to war on the horses that we want. He who puts up his hand with the nosebleed wins.' Tinkler was crystal-clear about his intentions: he was following the Woodlands blueprint. 'It took those guys [Jack and Bob Ingham] 25 years to build Woodlands,' Tinkler said, 'but we are trying to replicate it in three.' It was confidence bordering on folly: from this point on, many in the racing industry marked Tinkler down as naive in the extreme.

Tinkler's former employees have described 2008 as a spendathon. Remarkably, it was said that in those early auctions Tinkler was buying so many horses, through different agents, that he sometimes ended up bidding against himself.

Inevitably, buying in such a hurry created the potential for mistakes, and right from the beginning there was blowback about Tinkler overpaying. Referring to the oil-rich dad in the 1960s sitcom *The Beverley Hillbillies*, an item in the *Sunday Age* had one wit observing that there was 'something of the Jed Clampett about [Tinkler]'. The article predicted that he would go broke trying to found his racing empire, and that there'd be plenty of people willing to help him spend his dough: 'it's called redistribution of wealth'.

*

Hopes for Patinack were high. The racing industry in the first half of 2008 was still reeling from the outbreak of equine influenza the previous August, which had triggered a nationwide ban on racing and horse standstills in four states. The industry had been brought to its knees – at one point, it had even seemed that the Melbourne Cup would be delayed. The outbreak had cost hundreds of millions of dollars – but then Tinkler had appeared on the horizon, like a saviour, and rode a wave of gratitude.

Tinkler was feted: in April the gossip pages were aflutter after he was spotted at the Sydney Institution Bondi Icebergs, lunching with the Magic Millions' owners, John Singleton and Gerry Harvey. Singleton hosed down speculation of a deal with Tinkler, telling the *Newcastle Herald*: 'We talked about a lot of things, but selling any share of Magic Millions, or my Strawberry Hills stud, or Gerry's stud interests, were not mentioned. It was more a good day out talking about footy and whatever.' Suddenly, Tinkler was making rich and powerful friends.

Patinack's racing colours – blue/green silks and a gold cap – made a decent first impression on the racetrack. Two of its horses made the $2-million Golden Slipper at Rosehill, the richest race in the world for two-year-olds. Patinack had a couple more Group One winners over the next few months; Tinkler's then racing manager, Mark Webbey, later recalled that, from a standing start at Christmas, it had been 'a pretty impressive start to the Patinack brand'.

Stunningly, in May 2008 Woodlands was sold to Darley, the global racing operation run by Dubai's Sheikh Mohammed, for half a billion dollars. This galvanised Tinkler. He wanted to create a stud just as valuable.

In June that year, Tinkler put on a swish cocktail party on a Sunday night at the Gold Coast's Palazzo Versace, for 150 racing identities, heralding his entry to the sport of kings. The *Gold Coast Bulletin* billed the event – which was MC'd by newsreader Sandra

Sulley, with entertainment by comedian Vince Sorrenti and singer David Campbell – as a 'coming out parade', where the 'big man with the big wallet' let his guard down to explain his 'very Australian' vision for Patinack.

One of the event organisers, a Patinack employee given the fake name of 'David Cole' to protect his identity, later told ABC reporter Brendan King, of Radio National's *Background Briefing* program, that he had convinced the rather shy Tinkler into opening up a little for Sky Racing's Andrew Bensley:

> I got Nathan up on stage and Andrew interviewed him and got him to talk about his vision and where he was going and he was quite staggering, and we would have had 500 people at the Versace Hotel. We had comedians, we had entertainers, it was just a phenomenal night. People were blown away … everyone in the room was just absolutely silent as people hung on every word that he talked about. About the vision, about what his goals were, objectives … he was outstanding. It was fantastic. And our credibility was – well, the industry was so supportive, and he got a standing ovation when he was finished.

Just three weeks later, though, eyebrows were raised when Tinkler sacked Andrew 'Bowie' Bowcock and his wife, Lasca, who had stayed on as bloodstock managers at Patinack. They were highly respected in the racing industry, having built up one of the finest boutique studs in the country. Tinkler had bought it and then walked right over them.

Bowcock was bemused. 'For some unknown reason he sacked us,' he told Ben Hills much later. Even in his short time at Patinack, Bowcock started hearing about suppliers who weren't getting paid. 'Nathan was trading off our good name because no one knew him from a bar of soap. There was the feed, the farrier, the vet – he would just hang them out to dry until they went for a summons. It's just the way he is. "Because I can" should be his motto.' Bowcock moved to Queensland and now has his own bloodstock business on the Gold Coast.

On *Background Briefing*, 'David Cole' explained that Tinkler just didn't relate to the Bowcocks: 'He thought they were old school and he was young school. And that he didn't see them going the way that he wanted things to go. And he had not necessarily given them clear goals and objectives on how he wanted things done. They were the ones that knew the business and they were tracking along the way they would normally run their farm.'

Cole explained the heavy impact that slow payment of bills had on the smaller operators in the racing industry, like trainers, feed stores, and farriers.

> DAVID COLE: They very much rely on cash flow to keep their businesses running. They're much smaller, they don't have the funding behind them. Trainers in particular have a considerable immediate cost. You've got horses to feed and you've got track orders to pay and rent on stables and vet bills that just keep coming in. They require cash flow immediately, so the impact on them is enormous.
> BRENDAN KING: Do you think people are mistrusting of the Patinack brand?
> DAVID COLE: I don't think that they've done anything to instil the confidence in the brand.
> BRENDAN KING: David Cole says the industry's reluctance to accept Tinkler was of his own doing and not because thoroughbred breeding is an exclusive club.
> DAVID COLE: Look, I think whenever you've got an industry like that, it's not so much that there's a club – well, obviously there are very significant players and very significant breeders

that have been around for a long time, but they rely on external factors that enable them to sell at the right price. He was never going to be welcomed into that group with the ways that he was himself. He was non-communicative. He was single-minded in what he wanted to do. He couldn't demonstrate understanding, or be prepared to take the advice of people that were well-regarded that could help him establish himself. Once again, people like the Bowcocks. And that's what really demonstrated those issues that allowed him to demonstrate his inconsistency and created such uncertainty.

The Bowcocks' sacking was a portent. For now, however, the rumours about Tinkler's style would be confined to the horse-racing fraternity.

Tinkler's spending did not let up. In July 2008 Patinack caused a stir by nominating 240 unraced yearlings for entry into the following year's Golden Slipper, forking out $84,000 in entry fees – all in the hope that two of its horses might qualify to run. By comparison, established trainer David Hayes only made 92 nominations, Gai Waterhouse 75 and Lee Freedman 73.

At Japan's premier sales, Tinkler spent another $1.6 million, buying five yearlings and a foal, and made expansive comments about his ambitions in racing. He told reporters how his Japanese yearlings would be sent to Europe for training. His 'ultimate goal', he said, was 'to found a racing and breeding business including properties, racing stables and stallions across the globe ... we want to end up with stud operations in the northern and southern hemispheres. We're very keen on racing in Europe and Japan and the US, so we want to become a global player.'

He followed his spree in Japan with more buys at Deauville. This time, his presence was noticed. Supposedly new to the European bloodstock scene, the 'free-spending Australian mining magnate' paid over $1.4 million for one colt, then indicated that it would be left in France for training.

Back home, Patinack's cheque books were still open wide. Tinkler bought the 580-hectare Swettenham Stud in the Hunter Valley for another $8 million – it would serve as a breeding property for his 250 broodmares, while the Segenhoe Valley properties remained as a stallion complex. On he rolled.

A banner day for Patinack Farm, which had sponsored the Cameron Handicap in Newcastle, came on 17 September 2008, when it had a winner in its own race with Raheeb. Waiting around afterwards was a *Newcastle Herald* journalist, Neil Jameson, who penned some beautiful words about what Tinkler was coming to represent:

> Only relatives or penniless punters hang around for the post-race presentations which makes you wonder why a lone male, oozing blue-collar bonhomie and a bourbon glow, has draped himself over the fence to hang on every word of Nathan Tinkler's acceptance speech. Baby face beaming, the owner of Cameron Handicap winner Raheeb is saying I feel like a bit of an Indian giver, sponsoring the race and winning it, but it was certainly very exciting for us.
>
> The male leaning on the fence responds with a 'You go, dude' and the penny drops. Of course; this is the Australian dream in its buff simplicity. The fence-hanger identifies with Tinkler, a fellow traveller who just a few gear shifts ago was another minor cog in the Hunter's coal boom. And now, here he is, a racing magnate. This is not envy, this is pure empathy. Yeah, you go dude.

*

By the end of 2008, the turnover of Patinack's top people was becoming impossible to ignore. In September Tinkler had split with his original agent and managing director, Roger Langley; that had been handled well enough. Tinkler had thanked Langley and taken charge himself, saying he wanted to 'amplify' his involvement: 'this is my passion and I plan to oversee all areas of the business, both operational and strategic'.

Two months later, though, Tinkler had a nasty falling out with his high-profile trainer, Anthony Cummings. The story did not emerge until Cummings' private company, Something Fast, brought a District Court case against Patinack Farm in late 2009. The trainer alleged that Tinkler's stud owed him $167,000 in unpaid fees, including payment for three horses bought at the 2008 Karaka sales.

Patinack quickly filed a defence, denying it owed the money, as well as a counter-claim for a whopping $6.4 million in damages from Cummings. Tinkler alleged that Something Fast had received over $2 million in secret commissions and incentive fees from vendors of horses it had bought on Patinack's behalf at the Magic Millions in 2008. By now disappointed in the performance of the hundred-odd yearlings that had been bought for him, Tinkler also argued that Cummings had breached his duty to buy sound horses that he thought would eventually win Group One races. He was claiming the loss in value of 15 horses which Cummings had bought for Patinack and then allegedly trained so hard that they had broken down or been rendered lame, preventing them from earning prizemoney.

Cummings denied all the allegations: he had disclosed all his commissions, he had no such duty to buy winners, and he had not trained Patinack's horses until they were lame. It was clear there was no love lost between the two men, and now it was all-out war.

Issuing a blizzard of subpoenas, with only mixed success, Tinkler tried to obtain the tax records and bank statements of Cummings, his wife and all their companies, as well as lists of all their assets and liabilities, and the financial records of Coolmore Stud and the other vendors he suspected might have paid commissions to Cummings. Tinkler sought unsuccessfully to interrogate Cummings about an apparently unexplained increase in the value of his assets. According to one of his affidavits, but without any further evidence, Tinkler believed that 'at the time of his conversation with Mr Cummings about purchasing horses for him, he was aware that Mr Cummings was living in modest accommodation, but after the purchase of the horses subject of these proceedings … Mr Cummings moved into expensive, luxury accommodation'. Tinkler also subpoenaed Racing NSW to get every complaint made about Cummings since 2000.

Adding fuel to the fire was Tinkler's bitter resentment that Cummings had landed him with a dud stallion, Sidereus, which Patinack had bought for a hefty $2.5 million in April 2008. It would be some time before the dispute hit the courts, but the problem had become plain soon after the purchase: Sidereus was a 'rig' – meaning he had only one testicle.

Tinkler felt duped. Patinack had relied on Cummings but had not been told either that Sidereus was a rig, or that Cummings was a 10 per cent owner of the horse and would receive a commission on the sale. The purchase had been made in halves, with the second instalment due after Sidereus retired from racing. In 2010, believing Patinack had retired the horse, the vendors sued Patinack for the balance of the purchase price. Patinack counter-claimed, asking the court to void the whole transaction: either it was based on a common mistake, if the vendors were unaware Sidereus was a rig, or it was based on misleading and deceptive conduct, if the vendors were aware.

Tinkler claimed that Cummings had told him Sidereus was a 'great looking yearling that is developing really well'. Sidereus was not of merchantable quality as a stallion, and certainly not worth the purchase price; horses were not bought for $2.5 million unless they had two functioning testes and breeding prospects. Cummings admitted that Tinkler had not been told Sidereus was a rig, or that Cummings had a stake in the horse, but he denied that the horse was bought for breeding. His brief was to buy for Patinack a horse qualified to run in the Golden Slipper, Cummings claimed – and anyway, he denied Sidereus had been overpriced or was unfit for use as a stud horse. (Sidereus, later offloaded by Patinack, stands today as a stallion for a fee of $6000.)

The claims and counter-claims continued to mount, and the cases got increasingly tangled. Neither Tinkler nor Cummings would back off. Tinkler wanted to make an example of Cummings, while Cummings' lawyers accused Tinkler of a 'malicious personal attack'. Finally, in October 2011, days before the messy dispute was due back in the Supreme Court, Gerry Harvey brokered a confidential settlement deal over a four-hour dinner – seafood and pasta and a $58 bottle of Cloudy Bay sauvignon blanc – in a private room at Darcy's, in Paddington. The *Sydney Morning Herald* reported that Tinkler had agreed to pay Cummings about three-quarters of the $1.2 million he was claiming, but a Tinkler associate claimed the amount was much less.

It wasn't just the Bowcocks and Cummings whom Tinkler fell out with in Patinack's turbulent first year. The Gold Coast trainer Jamie Nielsen, who'd been effusive in his praise of Tinkler when recruited to Patinack early on, was next to be pushed aside. So was Tinkler's Victorian racing manager, Patrick Payne. He would soon add Kerry Packer's former lieutenant Peter Beer, trainer Rick Connolly, Mark Webbey and Don McKinnon to his growing list of former managers, many of whom were sacked without their entitlements being paid.

Mark Webbey was fired as Patinack's racing manager in late 2009. He later told *Background Briefing*:

> I was called into the office one day on a Thursday, and I was on my way to Wyong races to tell you the truth. We had runners there that day. And I got a phone call to say 'Be in the Randwick office by 1 o'clock' and I was asked to go upstairs to meet with Troy Palmer and I was dismissed on the spot.
> BRENDAN KING: You didn't see it coming?
> MARK WEBBEY: No, I didn't see it coming.
> BRENDAN KING: Were you given a reason?
> MARK WEBBEY: Nathan wanted to do everything all his own way. Nathan wanted to program his own horses, and my role was now redundant.

In early 2010, *Sydney Morning Herald* journalist Tom Reilly started digging into Patinack and began to uncover the real story behind Tinkler's racing operation – which by then had been dubbed 'Pat'n'Sack'. He reported on the treatment of Jason Coyle, a young Newcastle trainer whom Tinkler had promoted to take over from Cummings, and who had delivered a string of wins in early 2009, with Onemorenomore and Linky Dink claiming two of Australia's top five races for two-year-olds. Despite this success, Coyle was unceremoniously dumped by Tinkler within nine months.

Reilly also spoke to New Zealand trainer Roger James, who had been forced to put three of Tinkler's promising horses up for sale to recover his costs, after Patinack had failed to pay him for over six months. 'I had no problem with Tinkler personally but I couldn't run a business when I wasn't being paid for my services,' he told the *Herald*. 'It was a shame as the horses had ability but it seemed I wasn't going to get the money any other way. At the 11th hour Patinack came up with the money they owed and the auction was called off.'

Tinkler, of course, loathed the scrutiny he was coming under. When Reilly finally got through to him later that year, the junior mogul launched into a furious tirade that has been quoted many times since: 'You're a fucking deadbeat. People like me don't bother with fucking you. You climb out of bed every morning for your pathetic hundred grand a year – good luck.'

The contemptuous spray at Reilly showed Tinkler's true colours: having made a fortune worth hundreds of millions of dollars, he held nothing but contempt for the average wage earner. In fact, Tinkler had gone further. Reilly recalls that, after copping a stream of abuse, he challenged Tinkler to say where the 'Pat'n'sack' story was wrong. Tinkler answered: 'Mate, I'll tell you where it went wrong, it all went wrong when your old man stuck his fucking cock up your mother.' The newspaper was too civilised to print this, but the outburst prompted one witty editor to dub Tinkler the 'boganaire' – a word that appeared for the first time in the headline the next day.

*

Tinkler was indiscriminate: he didn't care who he offended. Partly, this was because he took a pretty dim view of the well-heeled, traditional racing fraternity. In early 2009 he threw his weight behind the push for a merger of the Australian Jockey Club – the country's oldest race club – and the Sydney Turf Club. His language was typically in-your-face, belying both his relative youth and the fact that he'd only been a player in the industry for 15 months. 'I think racing needs to have a good look at itself,' he expounded. 'I've got over $100 million invested in the racing industry so I'd be disappointed if the racing industry didn't want to hear my thoughts on the critical issues.'

Perhaps the race clubs didn't listen closely enough, as a few months later Tinkler penned a strident open letter in favour of the merger, which would not have made him many friends. 'Enough with the impassioned testimony from committees and members holding onto emotional debates,' he declared. 'Our much-loved industry is dying ... partner up and succeed together, or die alone.' Tinkler may well have been right – a merger to form the Australian Turf Club did take place, in 2011 – but his arrogance was galling.

Tinkler had set himself against the racing establishment. As he told the *AFR*'s Angus Grigg in 2010, 'if ASIC governed this industry, half of them would be in jail. It's criminal some of the shit that goes on.' Until it was settled, Tinkler's case against Cummings would be dressed up as part of a campaign to clean up the industry, taking on self-regulation and a lack of transparency, as well as common sales practices like 'buybacks', in which studs secretly bought back some of their own stallions at auction for artificially high prices, to prop up their value and give the impression of a demand for their services.

'Nobody listens to me,' Tinkler complained. 'I have got 25 years until I am fucking 60 and they tell me that's when I can have a say. Sticking it up the pricks is what it's all about. Standing there holding the trophy saying, "Fuck youse, don't youse hate this?"'

Tinkler also revealed that he'd been personally ripped off – 'shortened up by scumbags' – in 2008, during his first big foray into the sales: 'If it didn't cost me ten or 20 million [dollars] I'd be surprised, in overpaying for horses and that sort of stuff.' But Tinkler was adamant that he had a long-term strategy for Patinack, with three important stallion prospects looming in 2011. He wasn't interested in winning a few races; he wanted to dominate the racing industry. 'It's a lifetime investment,' he said. 'I won't get sick of this. The bastards won't beat me.' His spleen knew no bounds. One wag commented that Tinkler had 'chips on both shoulders'.

Grigg noticed a negative strain in the names Tinkler was giving some of his horses: there was 'Frustrating', 'Heard the Latest' and 'Small Minds'. But Tinkler was having some fun, too. 'Let's Burn a Debt' raised a few eyebrows. Endearingly, many of Patinack's horses were named after Tinkler's kids, with names such as 'Win for Wilson', 'Levi's Choice', 'Run for Layla' and 'Cuddles for Naara'.

The bad feeling between Tinkler and the industry was obviously mutual, as *Four Corners*' reporter Stephen Long discovered when he visited the Richmond Valley, where Tinkler had stables for a time. In the quiet hamlet of Clarendon – home to not much apart from the Hawkesbury Race Club, a pub and a couple of horse studs – the tranquillity of a summer evening in December 2010 was broken by an explosion and the sight of flames. At a horse stud behind the racecourse, a large horse float belonging to Patinack was burning, having apparently been torched. Police investigated but never found the culprit. In truth, it could have been almost anyone; there was no shortage of enmity for Tinkler.

One ex-Patinack employee contacted by *Four Corners* recounted how he had once tried to bury the hatchet with Tinkler, sending him a text message proposing coffee. The reply came back: *No thanks. Ur just another cunt that I give an opportunity to, wasn't up to it and dogged me everywhere so fuck off. I check the paper for your funeral notice you fuckn deadbeat.*

Examples of bullying behaviour are legion. One Patinack creditor who was owed $80,000 took Tinkler to court for breach of contract. After racking up a huge legal bill and months of stress, they got a phone call from Tinkler personally, who said he would 'spend a million dollars in court to make sure we don't get a cent'.

Making it worse for those creditors owed money by Patinack, Tinkler's brazen spending was continuing. At one stage he was buying a horse a day, every day. It defied all accounting; nobody could keep track. Tinkler himself wouldn't listen to advice, as David Cole explained to *Background Briefing*:

> DAVID COLE: Nathan's a big – he's a catalogue man. So we used to say whenever there was a sale and there was a catalogue out, he'd buy. He could go through a catalogue – horses always look good in catalogues, there's lots of black type and good form, and they're bred from well-related families – but when you see the thing in front of you, you think that thing's so small that it will never be able to race competitively. He just believed that a thousand dollars, it might have been off a stallion that cost $30,000, he just thought that was great value.
>
> BRENDAN KING: Did you suggest otherwise?
>
> DAVID COLE: Of course. He's quite his own man in that sense. I said to him, 'Look, it's your money. In my view they won't race, you're wasting your money on them; they're too small.' But he needed to learn that lesson. There were horses that I was saying and that others were saying, 'This would never get to the race track,' but he still wanted to go ahead and do that.
>
> BRENDAN KING: And you'd say that to him, would you?
>
> DAVID COLE: Oh yes. We used to fight like fury over – I would just call him every name under the sun, and that he was wasting his money buying that sort of horse. He used to say if you've got the money, then you can make the decisions. I own it, so I'll do it. That's the nature of the man.
>
> BRENDAN KING: Despite the feuds, David Cole says he valued his time working for Tinkler.
>
> DAVID COLE: I still say to people to this day, we had a terrific time, and I bear him no grudge. He was very good to me. We had a lot of fun, we had a lot of laughs. And I gave up a huge chunk of my time. This is a love job, I didn't make any money

out of it, I barely covered my time away. But he was giving me the opportunity to really get involved in the thoroughbred industry, and you know, I travelled the world and represented him on that basis. We had a lot of laughs and it was a good time. It was difficult because his major mining deals: there was a fair amount of turbulence around getting the deals done. If I was going to be critical of Tinkler, he always committed himself to his next purchase before he had the money. But he always came through on those.

At the end of 2008, newspaper reports put Tinkler's total horse-racing spend at more than $100 million; by the end of 2009, they were reporting $200 million. Tinkler himself made similar admissions, telling the *Daily Telegraph* in early 2010 that Patinack, which had 150 staff and 300 suppliers and a presence in four states, cost $20 million a year to run and he'd invested $200 million.

Figures from the website www.stallions.com.au, tracking Patinack's buying at public auctions, show that his biggest splurge was in 2008, when he spent $40 million on 192 horses – a tally which does not count the most expensive stallions, or horses bought privately or overseas. In 2009 that figure dropped to $13 million (another 78 horses), then it kicked up again in 2010 to $18 million (on 102 horses), then fell to $5 million in 2011 (on 35 horses). Finally, it trailed away in 2012, when he bought just two horses for less than half a million dollars. The total, over five years, was $77 million for 409 horses, with two-thirds of that number in Patinack's first two years. In addition, between 2008 and 2011 Tinkler paid at least $50 million for extensive stud farms in the Hunter Valley and the Gold Coast hinterland.

By now, there were serious questions about the return Tinkler was making on that investment. His winnings were modest. In September 2009 he appointed John Thompson – a long-time foreman for Bart Cummings – as Patinack's head trainer. Finally, Tinkler had found someone he could stick with, and Thompson would repay that loyalty. But he still had a hard time preparing winning horses at Patinack: over five seasons from 2009–10 until 2013–14, Thompson generated $20 million in prizemoney but only four Group One race wins, out of almost 4000 starts.

Some expensive horses, such as Metallurgical, proved disappointing: he would never win a Group One race, and his career winnings were just $286,000. Metallurgical – named after the coking coal at Middlemount that had created his fortune – cost Tinkler plenty over the years, as betting records later tendered in court showed. On a particularly bad day at Rosehill in late 2009, Tinkler plonked $200,000 down on the three-year-old – $100,000 for a win at $3.80, another $50,000 just 30 seconds later, and another $50,000 five minutes after that – only to watch him run second. All up, Tinkler dropped over $300,000 on Metallurgical, until finally he backed him to win on the Gold Coast in early 2011 at odds of 7–1, and collected $140,000.

In October 2010, at Melbourne's glamorous Derby Day, Tinkler almost achieved his dream of having a runner in every race, just like the Inghams. But his best hope for a winner on the day, Trusting, ran only eighth. The disappointments piled up. Tinkler's one-time racing manager Mark Webbey told *Background Briefing*:

> Nathan's very passionate. While we were winning, he was wonderful and supportive and very generous. But there were days when we did get beat, and it was difficult. The text messages were flying left, right and centre. But as I said, he's very passionate, wears his arse on his sleeve and we got to expect that if we didn't win a race on a Saturday, well we expected to be given a bit of a spray over it.
>
> BRENDAN KING: What did the text messages say?

MARK WEBBEY: 'It's not good enough, you're an embarrassment to the Patinack brand' and so forth. 'My horses are better than this' and you know, what we came to expect, a bit of volatility within.

Of course, Tinkler had no one but himself to blame. One industry observer memorably told the *Sydney Morning Herald*'s Ben Hills:

The horse-racing industry thinks he's a bit of a joke; he's haemorrhaging money like you wouldn't believe. Anywhere in Australia there's racing you'll find Tinkler has a horse and it's the most expensive horse in the race and it's running seventh. He is getting a terrible bang for his buck. He's hired and fired people, used them up and spat them out in a most brutal fashion. He's also got a trail of creditors you wouldn't believe.

*

For the amount of money Tinkler had put in, Patinack's stud income was also meagre. Figures compiled using *Australian Stud Book* data show that service fees paid to Patinack actually fell over the stud's first three years, even as more stallions were added to the roster. In 2008 Patinack stood Beautiful Crown (inherited from Alanbridge), Casino Prince, Husson and Wonderful World, for a total income of $9.9 million. In 2009 Murtajil was added to the roster, earning $2.2 million, but the stud's total service fees fell marginally, to $9.8 million. In 2010 Raheeb and Teranaba were added, but they earned very little and total income fell to $7.1 million. In the 2011 season, two foreign stallions – Lope de Vega, from Ireland, and Monaco Consul, from New Zealand – brought in an extra $3.7 million, boosting total income to $10.7 million. But in the 2012 season, assuming that 75 per cent of coverings produce a live foal, the stud's overall income remained steady at about $10.6 million, despite the addition of a tenth stallion, Trusting, who brought in $1.1 million. All up, in its first five seasons, Patinack generated roughly $48 million in stud fees.

Another way to cut the *Stud Book* data: of the top 120 sires in the 2012 season, by number of coverings, Patinack had six, theoretically producing gross income of $12 million at their quoted service fee (assuming, for the moment, every covering produces a foal). Darley, by comparison, would produce $83 million from its 21 stallions, while Coolmore would produce $87 million from its 10 stallions.

At the heart of Patinack was an unusual and risky strategy: regularly patronising the stud's broodmares with its own, unproven stallions. Successful commercial studs like Woodlands did this sparingly, perhaps only in 10 per cent of coverings. The *Herald*'s Tom Reilly worked out in 2011 that 58 per cent of the coverings by Tinkler's stallions had been from his own gargantuan band of 550 broodmares, which by then was the largest in the country by more than 100. Reilly wrote:

> Where Tinkler has saved money – at least in the short term – is by sending almost all his mares to be covered by his stallions, which means he has not handed money to rival farms. But this tactic is fraught with risk.
>
> Should his stallions turn out to be failures, he will not only lose money invested in them but will also send the value of his collection of mares, acquired for more than $70 million, plummeting.
>
> And if the horses he breeds are not up to winning the biggest races, not only is he wasting tens of millions on training fees, his chances of creating another money-spinning stallion are also greatly diminished.

> 'It's foolhardy in the extreme to breed all your horses to your own unproven stallion,' a leading agent says. 'Most stallions are ultimately deemed failures but, if one of his sires can hit home runs and somehow Tinkler's tactic works out, then he'll look like a genius.'

Tinkler had always maintained that it would take years to judge the success of what he was doing at Patinack. In early 2009 he told the *Daily Telegraph*: 'I don't think we'll be able to put our hands on our hearts and say this is really working for 5 or 10 years yet. When the Casino Princes, Wonderful Worlds and Hussons produce Group One winners for us standing at stud and we are replicating our own bloodlines, that's when I feel we will be successful.'

In 2009 Helsinge, at Gilgai Farm in Victoria, gave birth to a colt, All Too Hard, which had been sired by Patinack's Casino Prince. All Too Hard was a half-brother to the sensational mare Black Caviar and was trained by John Hawkes. Patinack bought All Too Hard at Inglis's yearling sales in Sydney in 2011, paying $1.025 million. The colt was to become Tinkler's great hope.

4.

SPENDATHON
BOYS AND THEIR TOYS

Horseracing wasn't the only way Tinkler found to spend his millions. After the Middlemount deal, it didn't take long for Newcastle's real-estate records to start tumbling.

In early 2008 rumours spread that Nathan Tinkler had bought what might be the city's grandest home, 'Jesmond House', on the top of The Hill, a few steep blocks from the CBD, where the original coal barons of the area had built mansions at the turn of the 19th century. ('Jesmond House', however, had been built by John Wood, the city's biggest brewer, who'd made a fortune selling beer to thirsty miners during the first coal rush.) As it turned out, it wasn't Tinkler but his erstwhile partner Matthew Higgins who had bought the historic mansion for a jaw-dropping $7 million – the highest price ever paid for a house in Newcastle.

Tinkler had his eyes on Merewether, the classic city beach with one of the most beautiful ocean baths in the country. A favourite jump-off point for hang-gliders, Merewether has cracking views in every direction and has been bombarded with luxury developments, morphing into Newcastle's dress circle, although it hasn't yet lost all its character.

The agents made the usual lame pretence at keeping it quiet, but Tinkler was soon connected with the February sale of rugby league legend Andrew Johns' home for $4.3 million – a record price for the suburb. The four-bedroom house, at 1 Berner Street, is right on Dixon Park, which slopes down to Bar Beach, and has big decks to soak up the sun and the views.

At this time the Tinklers were living in their Pullenvale mansion on the fringe of Brisbane, and the Merewether house was apparently to be used as a weekender. Tinkler didn't care what he paid. As one local real-estate agent told the *Good Weekend*'s Ben Hills: 'He was a very naive purchaser, like a Lotto winner. In this case the big red ball came down right on Andrew Johns' head.'

It was just the start, though. Within a month Tinkler had changed his mind, deciding that Merewether's most expensive home wasn't good enough. He wanted the even bigger pile on the other side of the park, at 4 Ocean Street, which faced south but had a vacant block next door, at number six; owned by the Steggles family, it had only a dilapidated 1920s cottage. Tinkler spent $8.8 million buying both properties.

Unhappy with the kids' schooling in Brisbane, Rebecca soon moved the family back to Newcastle, and they moved in at Ocean Street. Choosing high-profile architect Rosie Stollery, a feng shui expert from Byron Bay, they started drawing up plans to build Newcastle's most expensive home on the massive site – a $13-million, two-storey, six-bedroom mansion with a basement garage and cellar, courtyard pool and pavilion. The development application was lodged in 2011 and was approved, despite a handful of objections from neighbours who lost their views of the beach.

One drawback with both Merewether properties, however, was their complete lack of privacy: hundreds of people walk within metres of the exposed decks and living areas. At the Ocean Street house, which was overlooked by a block of units, anyone could walk right up to the side windows for a stickybeak.

The Tinklers decided they needed a retreat. In October 2008 they bought another mansion, far from the crowds this time, at Sapphire Beach, just north of Coffs Harbour. An amalgamation of three home sites, 'Noorinya' has a private beach, tucked between the Opal Cove golf resort and the south end of Campbell's Beach, where half-a-dozen other concrete-and-glass getaways jostle for 10 metres of absolute beachfront, deck chairs and pool fences cramming to touch their own tiny patch of sand.

'Noorinya', say those who've been there, is pretty classy. Built by Microsoft pioneer Jaybe Ammons, who spent something like $2.8 million amalgamating the site a decade earlier, the home set Tinkler back a cool $11.5 million – a new record for the mid-north coast of New South Wales. A Balinese-style compound with a series of conjoined pavilions sprawling over four and a half acres of land, surrounded by remnant rainforest, 'Noorinya' was perfect for Tinkler, being just 10 minutes from Coffs Harbour's airport, and roughly halfway between Brisbane and Sydney.

'Noorinya' had the lot: a discreet entrance off Coachman's Close, a long driveway, an open-plan home with three kitchens (including a butler's pantry), a cinema, a gym, two offices, a three-car garage, a 20-metre lap pool and a lazy inclinator down to the sand. Its marketing clip (still viewable on YouTube) was suitably extravagant:

> There is a place across the ocean that dreams are made of. A place of seclusion and serenity, where you'll wake up to the sounds of the sea. Where you'll find dolphins at your doorstep, and make footprints in the sand of your very own beach ...

For the next few years, 'Noorinya' would be Tinkler's trophy home, the place he'd invite business contacts he wanted to impress.

*

Having now spent more than $30 million on four mansions in Brisbane, Newcastle and Coffs Harbour, Tinkler needed a jet to fly between them. The private jet is the mark of the super-rich, the epitome of the high life, the ultimate boy's toy. So, in December 2008, Tinkler bought his first plane, a brand-new twin-engine Hawker 400XP, for around $6.5 million. Now he had really made it.

Capable of flying at 866 kilometres per hour, the Hawker 400XP had leather seats for seven passengers and a range of more than 3000 kilometres – perfect for flying up and down Australia's east coast. But Tinkler was a big guy, and with a cabin area of just 10 cubic metres he soon found the plane too small for his comfort. In a repeat of his Merewether spree, he decided he needed a bigger one, and upgraded to a $13-million Hawker 850XP, which was no faster but could seat a dozen passengers and had a range of 5000 kilometres, enough to fly almost anywhere in Australia nonstop. And it had plenty of room for Tinkler. For the next couple of years, this plane would be his workhorse.

Beyond spending fistfuls of money, however, very little constructive work was going on. To be fair, the second half of 2008 saw the peak of the global financial crisis. When Lehman Brothers collapsed in September that year, credit markets froze worldwide, interest rates spiked and any company stuck with too much debt was taken to the brink. On the flipside, there was panic selling taking place everywhere, and Tinkler had entered the crisis with hundreds of millions of dollars and very little debt. Cash was king, and so the GFC could have been a very happy hunting ground for Tinkler.

His mentor Ken Talbot, for example, with just a small team of 'geos' based in Brisbane, won out with a string of speculative mining investments after selling down his stake in Macarthur Coal. These included the Mozambique coal project Riversdale Mining (before it was sold to Rio Tinto for a bomb), the iron ore explorer Sundance Resources, and the offshore oil and gas explorer Karoon Gas, all of which were, at least for a time, spectacularly successful.

Whatever trouble he'd gotten into with the corruption inquiry in Queensland, Talbot was a brilliant miner, and for a while it seemed that everything he touched turned to gold. By contrast, almost without exception, Tinkler's investment decisions in this period went bad. So much for his legendary ability to pick undervalued mining assets.

Tinkler had created Aston Resources in January 2008 and Boardwalk Resources in April, but while these two coal exploration companies would later play a key part in his fortunes, they did very little in this period apart from early work on a farm-in deal with listed junior Cockatoo Coal on the Dingo project in Queensland. Announced in April 2009, the deal gave Aston up to 70 per cent of three tenements in the Bowen Basin – close to but separate from the Dingo West permit pegged by Custom Mining and sold to Macarthur in late 2007 – if it spent up to $9 million on exploration over 27 months. Progress through 2009 was fitful and the project was later rolled into Boardwalk.

Tinkler's moves away from coal exploration proved a mistake. In 2008 he bought short-lived stakes in a string of junior miners, including $2.3 million for 15 per cent of the listed base metal and uranium explorer Mt Isa Metals; an unknown amount (probably more than $5 million) for a 7.5 per cent stake in the South American Iron and Steel Corporation, which had interests in Ecuador and Chile (and turned out to be a dog); and some millions wasted on the ill-fated CopperCo, a Queensland explorer which collapsed under a mountain of debt in December 2008 amid controversy over woeful disclosure and huge volatility in the copper price. Investors lost the lot.

At the end of 2008, Tinkler hired a new chief executive for Aston Resources, the qualified mining engineer Hamish Collins, who had worked in corporate finance in the resources industry for a decade. Aston Copper was established in December, and Collins put his foot on a viable project: the remote Walford Creek, a low-grade copper and zinc deposit north of Mount Isa. Walford Creek, however, took years to progress and would never deliver a return to Tinkler.

Worse still, Tinkler started punting on the stock exchange like an ordinary day trader. Convinced of his genius in the mining game, between 2008 and 2009 Tinkler took out huge margin loans against stakes in BHP and Rio, whose shares were beset by the global financial crisis. Anyone familiar with margin-loan investing knows that it's not for the faint-hearted: if the market turns, it is easy not only to lose the principal you invested but also to end up with massive negative equity – to be on the hook for the entire amount you borrowed, but with no asset to set against it. It seems incredible, but former insiders swear that $70 million to $80 million went up in smoke as Tinkler gambled on mining stocks.

Compounding Tinkler's financial problems was the small matter of an unpaid capital gains tax bill from the sale of his shares in Macarthur Coal. Notionally, at the flat company tax rate of 30 per cent, having contributed less than a million dollars in equity, Tinkler Investments was liable for more than $130 million. Tinkler held out as long as possible but eventually agreed to pay off a confidential amount in instalments; on oath years later, he said it was $100 million.

Tinkler still wanted to expand into property development, and in November 2008 he agreed to pay $8 million for a 10 per cent controlling stake in one of Newcastle's most successful businesses, the Buildev Group (the deal was structured to minimise stamp duty but actually gave Tinkler a majority interest). Built up by founder David Sharpe, the company had a string of successful projects across regional Queensland and New South Wales, specialising in light-industrial and retail property. Buildev had just put its foot on a part of the former BHP steelworks site in Mayfield, in

Newcastle, where it successfully tendered to build an intermodal hub for the planned container port. It would soon become one of Tinkler's biggest headaches.

*

One of the investments Tinkler made during his spending spree turned sour very quickly, not only costing him millions but also attracting his first real negative publicity. In September 2008 he met the British expat and entrepreneur Tim Sommers, who had founded a new business called the Supercar Club. Tinkler, a petrolhead like his father from way back, joined the club at the recommendation of his bodyguard and personal trainer, Corey Baldock – a man Sommers would later describe as Tinkler's 'paid best friend'.

The idea of the club was simple: charge a bunch of rich guys an annual membership fee of some $40,000, which would let them get their hands on a mouth-watering list of exotic sports cars: from a $2-million, 16-cylinder Bugatti Veyron to an Aston Martin DB9 V12 or one of three Lamborghini Gallardos, or one of a string of Ferraris, Porsches, Bentleys and so on. Tinkler loved the club so much that he wanted to buy it.

Sommers had launched the club in late 2006, with a base at Sydney's Jones Bay Wharf, and soon attracted a fairly A-list membership of roughly a hundred people: Mark Webber, Mick Doohan, Jason Akermanis, Jennifer Hawkins, James Morrison, James Packer, Ryan Stokes and Mark Philippoussis were all supporters in one way or another. The Supercar Club held race days at circuits such as Eastern Creek and Phillip Island – and even at businessman Dean Wills' famous private road, 'The Farm', near Gosford.

At first, Tinkler and Sommers hit it off. In an ebook he later published, *The Rise and Fall of the Australian Supercar Club*, Sommers recalled first meeting Tinkler in the foyer of the Westin Hotel, describing him as:

> A likable ... no really (!), humble bloke, as it then seemed. But very, very wealthy. $441 Million at the time I met him. At the time the downturn was in full effect, but Nathan was bullish. He vowed he'd leave the downturn with multi billions. Over several beers, Nathan ('Call me Tink; all my mates do') and I chatted about cars and motorbikes. I liked what I heard; passionate, enthusiastic, funny and very, very bright. Maybe not bright; maybe sharp. He had a good collection of cars and bikes and he spoke really passionately, and with great understanding, about his 997 Turbo, his Ducatis, Flying Spur, Jaguar XJR, Mercedes Benz E63 and Ferrari 612.

Tinkler sent his trusted finance chief Tom Todd over to do the due diligence, and within a few months he agreed to pay $2.2 million for a 49 per cent stake in the club, and to provide a limited-recourse finance facility, backed by the Tinkler Group, which would halve the club's monthly vehicle lease payments. Tinkler took over as chairman of the company, aiming to lift the club's membership and take it global.

According to Sommers, it only took a few months for things to go wrong, as he and Tinkler vied for control. Tinkler would phone to say he'd just seen one of the club's Aston Martins doing a burnout in Brisbane; it turned out that the club's car was actually in Melbourne. Sommers was shocked to discover that Tinkler had unilaterally shelled out $740,000 of club money to buy a Rolls Royce Phantom Coupe Drophead – a luxury car, not a supercar, and so missing the whole point of the club. Few members wanted to drive it.

Sommers put these incidents down to 'Nathan being Nathan', but as founder of the club, he had his own strong views and would not be pushed around. In his ebook, he writes that Tinkler was a 'control freak' who wanted to run the club his way – 'the wrong way', as Sommers saw it. 'Nathan's wealth hasn't come from running a business,' he wrote. 'It's come from speculating (call it gambling if you like), so he has little idea of how to manage a profitable business.'

Tinkler's uncouth side, funny at first, started to grate on Sommers. One time Tinkler left over $100 in notes and coins in the glove box of the club's Merc, telling Sommers: 'I don't need that I have plenty more.' Another time Tinkler bought *two* Desmosedici superbikes – worth $134,000 each – because he couldn't decide which colour he liked. At one point Tinkler decided that members of the Australian Muscle Car Club couldn't join the Supercar Club; 'we don't want those bogan cunts,' he said. Tinkler, a Ford man himself, was clearly now putting his humble origins behind him.

Tinkler also boasted of being 'an expert in great cars and fine women'. One time Sommers emailed Tinkler a picture of two men at a club event and received this reply: 'Where are all the topless chicks? We are male fukn chauvinists mate … how are we going to attract members?'

After a while, even the politically incorrect Sommers – whose ebook is itself full of asides such as how the Lamborghini SuperLeggera was known as the 'Superleg-opener' – found Tinkler's behaviour intolerable. His swaggering rich-guy charm had worn off. After first thinking Tinkler to be an eccentric 'man's man', he later realised his mistake, writing:

> Eccentric doesn't really describe Nathan well enough. At every level of business, the man-mountain is determined to, and beyond, the point of ruthless. He loves the skulduggery. I suppose it was inevitable. From being an electrician, to having hundreds of millions of dollars within such a short space of

time, he believes he's touched. Unfortunately it's not just the Midas touch. Amongst his other qualities he's assumptive and paranoid. Hell, one day I got 15 calls and 10 texts in which, I swear, he cursed every word in the alphabet. The man is very difficult when he doesn't get his own way.

The developing personality clash between the foul-mouthed Aussie and the slick Pom was about to turn ugly.

Sommers became increasingly concerned that Tinkler was failing to fulfil his side of their share-purchase agreement: Tinkler had not yet put in the full amount he'd agreed to, having deposited only $1.4 million. He had also failed to provide the half-a-million-dollar loan facility for the club's working capital by the end of June, as agreed.

For his part, Tinkler wanted control of the club's accounts. Without asking Sommers, he tried to put one of his own finance guys, an old schoolfriend, Richard Glenn, in charge. Sommers and his wife, who was also involved in the business, resisted. This fuelled suspicion between the business partners, and amid the general tussle for control of the organisation's direction there were soon accusations flying that Sommers was taking money off the club.

The acrimony worsened through July. Sommers believed Tinkler was starving the club of funds and demanded he put more money in. Tinkler criticised Sommers for getting out the 'begging bowl' and threatened to get his money back. 'Where's my $2.2 million?' he asked Sommers at one point. 'If this business needs $2 mil every six months, then I am fucking out.'

A withdrawal of $180,000 cash by Sommers in mid-August appears to have been the last straw for Tinkler, and on the 25th he tried to call a meeting of the Supercar Club's directors to stand Sommers down, pending a statutory audit of the accounts. Sommers found out that Tinkler had told staff he and his wife had stolen

$700,000 from the club and were not to be trusted. Sommers cancelled the appointment of independent director Peter Dempsey, who was siding with Tinkler, and boycotted the directors' meeting the next day, which was therefore invalid for lack of a quorum, and had the locks changed at the club's headquarters. He wrote invoices for the sale to himself of two Audis (worth $155,000 but sold for $90,000), and took them home. He also kept possession of four club motorbikes worth a total of $34,000.

With Dempsey's help, Tinkler changed the locks back the next day and told all staff that Sommers was gone, and he was not permitted to contact the club's employees, members or suppliers. Sommers had lost control of his business. Now under Tinkler's full control, Supercar sued Sommers a couple of days later to get back the cars and bikes, and the whole affair soon descended into another messy legal stoush.

On top of the allegations of misappropriation, Tinkler sued to get his own money back, alleging that Sommers had made misrepresentations, including that the Supercar Club was profitable and that there was no litigation against the business. It turned out that, unbeknown to Tinkler, Sommers had been sued by a UK company called P1, which was founded by the former Formula 1 driver Damon Hill; it was a forerunner of the Supercar Club concept. Sommers was to have established an Australian affiliate and had originally used the P1 brand, but had split with the parent in 2007 – taking intellectual property and reneging on a deal, P1 claimed – and had been sued for more than half a million dollars, which he'd paid. Tinkler suspected – wrongly, as it turned out – that the money he had put into Sommers' business had been used to pay off P1.

Sommers denied the allegations but left the country, which didn't help his case in the court of public opinion. He filed a counter-claim that the other directors – Tinkler, Dempsey and Todd – had engaged in conduct that was oppressive of him as a shareholder, and that he was still owed money for the purchase of his share in the club.

Meanwhile, although Tinkler assured the club's members of his long-term commitment, and sent an email telling 6000 customers how the Supercar Club would thrive without Sommers, its financial position was deteriorating. Sommers later wrote that Tinkler failed to buy a single new car, sell an old car or recruit any new members. The mismanagement under Tinkler's control was 'extraordinary', he wrote, and the club's revenue and membership numbers plunged by 80 per cent. Nine months after Tinkler seized control of the club, the whole thing was wound up.

When the Supercar Club case got to court and the men's affidavits went on the public record, they made for sensational reading. The media were obsessed by the case, and Tinkler came out all guns blazing, telling the *Sunday Telegraph*'s Paul Barry that he had been 'duped by a conman and defrauded of $4 million', and that more than 100 Australians had also been ripped off in the process.

Barry revealed that Sommers, a former British soldier, had been lucky to escape a jail sentence in the United Kingdom in 2002, after his computer company collapsed. He had incurred a three-year ban from serving as a company director, and had been fined £8500 for not keeping accounts and withholding documents. What's more, another of Sommers' companies in Australia had been wound up in 2008 because of its $450,000 tax debt.

It looked like Sommers had done Tinkler over. Tinkler's spokesman talked about the regrettable impact on the club's members: 'as an act of good faith to the customers … we are continuing to support the company financially,' he said, 'and hope that justice will be done as quickly as possible.'

When the case came to trial, however, the judge didn't see it that way at all. Tinkler, true to form, was a no-show. So was Tom Todd, who was genuinely overseas on business. In the Supreme

Court of New South Wales, Justice Richard White noted that there was no explanation for Tinkler's failure to give evidence, and that 'it can be inferred that his evidence would not have been of assistance to Tinkler Group'. Todd's explanation for his unavailability for cross-examination was 'unsatisfactory', the judge said, and his affidavit was rejected.

Sommers, on the other hand, turned up. The judge had 'serious reservations' about his credibility, saying he saw 'everything through the prism of self-interest', but noted that this did not mean all his evidence was rejected. In fact, Justice White accepted much of it. Yes, Sommers was still owed money: some $218,500, plus interest. And no, there had been no misappropriation. But Sommers should not have sold club cars and bikes to himself below value. Yes, Sommers had been oppressed by Tinkler and co. But no, Sommers should not have caused the club to take out a lease on a holiday home at Yarramalong, for the use of himself and his wife. Yes, Sommers had lied about the P1 litigation. But no, he was not liable for damages, because he had settled separately with P1 and there was no impact on Tinkler.

After hoping to get back his entire investment of $2.2 million, Tinkler was awarded nominal damages of just $10. In the event, Justice White's 112-page judgement, handed down in April 2011, did not reflect well on Tinkler at all:

> In my view Mr Tinkler and Tinkler Group conducted the affairs of [Supercar Club] in a way that was oppressive to and unfairly discriminatory of Mr Sommers. Notwithstanding that he had been the founder of the club and was a 49 per cent shareholder, he was excluded from management before Mr Tinkler had established whether he had misappropriated funds. In fact he had not. Whilst Mr and Mrs Sommers' delays in providing bank statements and financial informa-

tion were a contributing cause of Mr Tinkler's pre-emptive action, another contributing cause was the falling out between them. That falling out was partly due to Mr Tinkler's having made unilateral decisions in the management of [the club]. Another factor was Mr Sommers' having complained about Tinkler Group's not providing the working capital facility, and his invoking agreed procedures to attempt to resolve that dispute.

It would get even more salacious, as Tinkler fought desperately in court to suppress publication of Sommers' ebook, which threatened to rehash the material in his sensational affidavit, parts of which had been censored by the court. He obtained a court order that restrained Sommers from publishing any imputation that Tinkler was 'racist', 'chauvinist', 'does not pay his debts' or 'engages in immoral, inappropriate or unfaithful sexual relations'. The order was allowed to lapse when Sommers agreed to let Tinkler see the book at least 48 hours before publication.

*

It was no wonder that Tinkler was sensitive about accusations of infidelity. In September 2008, the *Sunday Telegraph*'s top gossip columnist, Ros Reines, penned an unflattering item about Tinkler. Its headline – 'Whale Tales' – alluded to one of Tinkler's nicknames, both for the money he was spending on horses and for his own girth.

> The hottest story in racing has been hotly denied by those involved. Millionaire breeding identity Nathan Tinkler, 32, whose fortune was only last week estimated by *BRW* at $441 million, has allegedly been rolling in the hay. The former mining electrician turned coalmine owner – and the richest

Australian under 40 – is rumoured to have been involved with a younger woman for nearly a year and his wife, a director of his companies, is seeking a divorce. Rebecca and the woman both strongly denied the rumours yesterday. Rebecca gave birth to the couple's fourth child at the end of August. This was about the same time that the alleged mistress was reportedly spotted driving into work in a flashy Lexus sports car, worth more than $100,000, instead of her old Toyota. Soon after, Rebecca Tinkler is believed to have commenced divorce proceedings, we're told to keep a hold of her share of the business.

Scalded, Tinkler went to ground, brushing off journalists at the Melbourne Cup. It wasn't until the next February that an item headed 'Still Together' appeared in the *Courier-Mail*'s gossip column:

> The richest of the nation's young rich, Nathan Tinkler, has set the record straight. Rumours had been swirling that the 32-year-old mining and racing figure … had split from wife Rebecca. 'Absolutely not,' Nathan told Confidential. 'Not true.'

*

For all the bitterness of the Supercar Club dispute, Sommers' ebook also provides glimpses of the high life Tinkler was enjoying, which was about more than the pleasure of putting the pedal to the metal in some of the world's fastest cars.

Tinkler and Sommers were about to launch the Jet Club, expanding the Supercar timeshare concept to aviation. Tinkler was planning to transfer his squashy Hawker 400XP over to the club; for a time he let pretty much anyone use it. He once told Sommers that the club could have it: 'I don't want the fucking thing anyway. I paid too much for it and I don't like it. Once it had someone else in it I didn't want it any more. It's like someone fucking your missus.'

In a fonder memory, Sommers told of one trip from the Gold Coast to Bathurst in early 2009, when Tinkler's jet was deployed to take legendary racer Dick Johnson and his wife to see his son Steven Johnson and Tinkler race a Ford Performance Vehicles GT Falcon in the Bathurst 12-hour race. On a stormy summers day, as Sommers recalled:

> … there was no mist on the runway at Coolangatta. I could quite clearly see the Hawker 400 XP – the ultimate in have-to-have boy's toys, lurking on the apron! Decked out in its livery of white with Regal Blue and Clarette striping, the 400 XP is more function than form … Once inside, it's as if I've been transported to Nirvana … little touches such as the seven leather seats arranged in 'center-club' format so that you can have a meeting without having to rearrange your Armani. And just to make you feel at home in this flying boardroom, you have wood panelling that extends down the length of the cabin and doubles as an armrest. All very comfy … we lift off the runway on our way to our cruising altitude of 35,000 feet … unbelievably this arrives in about 18 minutes, long before my first beer, and long after we enter a thick bank of cloud. In any of the smaller 'craft, these weather conditions would have me looking for that horrible little bag, but in the 400 XP there's only a gentle buffeting as we navigate the turbulence … The descent through the clouds into Bathurst is quite a bit bumpier than the ascent, obviously something to do with currents off the mountain. Again I marvel at the competence of the 400XP: for a relatively small aircraft it's an amazingly stable platform in these conditions … Sure beats Qantas hands down!

Steven Johnson and Tinkler ran 30th that year, beating plenty of did-not-finishers, and beginning what would become a pretty respectable partnership and friendship between a professional driver and a cashed-up 'gentleman racer'.

In late 2010, Tinkler even went close to buying a half-share in Dick Johnson Racing's storied V8 Supercar team, from Melbourne businessman Charlie Schwerkolt. Johnson and Schwerkolt were no longer on speaking terms, and the team was underfunded in 2010, dipping into its budget for future years just to keep going. In his autobiography, Dick Johnson recalled his first encounter with Tinkler:

> I first met Tinkler in a hotel car park at Bathurst ... it was a hot morning and he was about to strap down in a GT Falcon and race the 12-hour drive with my son, Steve. 'G'day Dick,' he said. 'Bloody good to meet you.' He was wearing thongs, shorts and a T-shirt. I had no idea who he was or what he was worth at the time. All I knew was that he was running a GT at Bathurst and he was obviously wealthy enough to support a race team and had a passion for motorsport ... I ended up having dinner with the man, and it turned out he was a good bloke who loved Newcastle, rugby league and his cars. [He] absolutely loved his Fords. His dream was to drive, knocking about in the lower categories. Tinks had the money to do it and I applauded him for having a crack. I found Tinks to be straight up and down, what you see is what you get.

In later meetings, Johnson broached the subject of Tinkler buying Schwerkolt out. Tinkler was interested. Johnson considered Tinkler the perfect potential business partner. After inspecting the team's Gold Coast factory and examining the books, Tinkler was on board. Two days out from the Bathurst 1000, which Johnson was hoping to win, news broke that Johnson's driver James Courtney and half his team would walk out at the end of the year. Johnson was shattered, as his book records:

> Nathan Tinkler sat in his private helicopter as the engines warmed. He had a cheque for $1.5 million in one hand and a copy of the *Daily Telegraph* in the other. He grabbed the paper and flipped to the back page. 'Turn the engines off,' he said. 'We aren't going to Bathurst.' Tinkler had booked a suite in a hotel in Mount Panorama and was due to arrive at Bathurst on the Friday. He was coming down to pay Charlie out and buy a half-stake in my team. All that changed in a moment, when he picked up the paper and read about the shit that was going on in the team. I was screwed. Tinkler never called that morning. I tried to reach him on his mobile, but it went straight to message bank. Deep down, even before I had gathered the truth, I knew why, but I was hoping he had been delayed or called away on other business.

Tinkler had had second thoughts; although a lower price was discussed, the deal ultimately went nowhere. Dick Johnson Racing battled on.

After racing their Ford for a few years, Tinkler and Steven Johnson switched over to a Porsche 997 GT3 after the rules changed so that non-production cars could race. Tinkler had to have his stomach stapled so he could lose enough weight to fit into the car's tight bucket seats, and comply with the Confederation of Australian Motor Sports' rules requiring a body-mass index under 30. (Tinkler's people denied the operation, but media outlets were soon reporting his remarkable loss of some 30 kilograms; his spokesman began handing out pictures of a much healthier-looking Tinkler, asking newspapers to stop using the old ones.) Tinkler and Johnson finished a very creditable fourth in the 12-hour, topping their class, after a memorable race in tricky conditions. Tinkler spun the Porsche at one point, when a spot of rain swept the mountain. A bit of fun, a shaken Tinkler told journalists. Tinkler and Johnson raced the Porsche on and off in the Carrera Cup for the next couple of years, generally finishing just off the podium.

Tinkler certainly was living life in the fast lane. In his 2010 interview with *Master Electrician*, the 'confessed revhead' ran through some of the favourite supercars in his garage:

Ferrari California – 'Practical and corners like it's on rails.'
Bentley Flying Spur – 'No such thing as a bad Bentley, great for that long drive, by yourself.'
Porsche 911 Turbo – 'Takes me two days to stop smiling every time I get out of it. Unbelievable power.'
GT40 – 'Fantastic car, until you want to see out of it or park it!'
Audi RS6 Wagon – 'Wolf in sheep's clothing, ultimate surprise package courtesy of Lambo V10. Looks like a Mazda 6!'
Mercedes Benz E63 – 'You will run out of road before you get to the bottom of this engine, it just keeps giving.'
Falcon GTP – 'Great value car, even better supercharged!! Not one for the family.'
Aston Martin DB9 – 'James Bond did not drive a car this unreliable? Gave it to my dad.'
Jaguar XKR – 'Great fun and does everything well. Looks fantastic.'
Ferrari Scuderia F430 – 'Should come with a kidney belt. I think they forgot to put the suspension in mine? But my god that engine! F1 on the highway!'

Tinkler loved his cars so much he started naming his companies after them: there was Zagato (an Aston Martin), McLaren, Veyron (a Bugatti), Cayenne (a Porsche), and Azure, Arnage and Mulsanne (all Bentley models).

Tinkler proved that no amount of money is so large that it can't be blown. As he told one journo: 'I don't owe anybody anything and I don't mind spending it.'

No one but Tinkler will ever know exactly how he chewed through the $440-million fortune he made out of the Middlemount deal over the 18 months to the end of 2009, but the tally would include something like $150 million to $200 million on Patinack Farm, $100 million in capital gains tax on the sale of his Macarthur Coal shares, as much as $100 million on questionable mining and property investments, including his failed margin loan punts, and at least $50 million on prestige property, private aircraft and luxury cars. It is not a widely appreciated fact, but by late 2009 Tinkler was almost clean out of cash.

Sadly, amid all his spending, much of it wasteful, what Tinkler did not do was pay back the people he owed money to, from the tough times before he got lucky.

5.

MAULES CREEK
LIGHTNING STRIKES TWICE

By early 2009, the global financial crisis had brought global mining giant Rio Tinto to its knees. It had fought off takeover bids by BHP Billiton and Chinalco, laid off more than 10,000 workers and frantically sold assets to pay down almost $50 billion in debt accumulated during its disastrous Alcan acquisition in 2007.

Having worked for Rio, Tinkler had a healthy disrespect for the company which bordered on contempt. No doubt he thought he could run its coal operations better than it could. At one point he tried to do just that, canvassing a complete takeover of their majority-owned New South Wales coal business, Coal & Allied, backed by the financial engineering whizzes at merchant bank Babcock & Brown. Talks were held with Rio but the idea went nowhere. Babcock itself would not survive the GFC.

Much has been made of Tinkler's ability to 'pick' undervalued coal assets, but the project he wound up buying off Rio in 2009 wasn't the one he first wanted. In October 2008 Tinkler had flown to London to meet Rio's then energy chief, Preston Chiaro, to make an offer on the Valeria deposit in Queensland, a massive thermal coal play that remains undeveloped to this day. Valeria was not for sale, came the reply from Rio, but would Tinkler be interested in buying the Maules Creek tenement in the Gunnedah Basin?

Maules Creek was a sizeable deposit containing predominantly thermal coal. It had sat undeveloped for decades, although it was only 16 kilometres from an existing train line between Boggabri to Newcastle.

Tinkler was reportedly 'stunned' to find that the project was for sale. Convinced he could extract value where Rio could not, he was immediately keen to buy it. To get the coal out from Maules Creek, most people were assuming that a new rail line would need to be built through the Liverpool Ranges, at huge cost. Tinkler's team dealt extensively with the Australian Rail Track Corporation and became confident a new line would not be needed. They would be proved right.

Rio put out an information memorandum in March 2009, which estimated that the Maules Creek deposit had a recoverable resource of some 398 million tonnes of coal – at least twice the size of Middlemount, although the offer documents did not estimate its proven reserves. It was unusual for a company Rio's size to run its own sale process. Ordinarily, investment banks would line up in a beauty pageant, pitching for the sales job and the chance to charge a fat fee. But these were not ordinary times, and Rio was moving fast. There was no time to re-evaluate the geological data on Maules Creek. Binding bids were sought by June.

Tinkler also phoned his old lawyer, Philip Christensen, who was holidaying in Bali when he rang. The two men hadn't spoken since their bust-up over Christensen's legal fees on the Middlemount deal. Now Tinkler apologised. He'd been under enormous stress. He'd got it wrong. He explained about Maules Creek, and how it was going to be the making of him all over again. He wanted Christensen back by his side, this time on the payroll, and offered him equity and the position of managing director of a new entity, Boardwalk Resources, which he'd set up to buy Queensland exploration assets.

Christensen felt that Tinkler had learned some lessons: in effect, by making the approach, he was asking Christensen to save him from himself. Christensen decided he didn't want to be a partner in a law firm forever. Within a few months he would quit Freehills, after 30 years, and work with Tinkler once again.

With the credit markets completely frozen, 2009 was a great year to buy almost anything if you were cashed up. That was as true of coal tenements as of any other class of asset. Coal prices plunged from their all-time peaks in 2008 as the GFC roiled commodity markets. Thermal coal was way off, back down around US$60 a tonne, and coking coal was down to US$150 a tonne.

There were other bidders for Maules Creek – including Xstrata and Macarthur Coal – but they had their own financial problems. Aston's stiffest competition would likely come from Whitehaven Coal, a specialist in the Gunnedah Basin which ran the nearby Tarrawonga and Werris Creek mines. Whitehaven had some cash and plenty of motivation, making it a natural buyer.

The company was led by managing director Tony Haggarty, a coal veteran who had made his fortune building up and selling Excel Coal to Peabody for $2 billion in 2006. Haggarty knew how to develop a coal mine and knew the value of an asset – at least, he thought he did. Whitehaven put in a bid for Maules Creek, believed to be between $250 million and $350 million, and was very surprised when it was knocked out of the running.

Aston Resources low-balled at first, putting in a bid of between $200 million and $250 million, but when Tinkler was told he was not going to make the cut, he immediately shot back: 'What's your number?' Rio wanted at least $500 million; Tinkler offered $480 million, subject to finance. He was in. In the event, Aston was at least $100 million higher than its nearest rival. It was Aston first, daylight second. Rio decided to give Tinkler a chance.

Tinkler was not nearly as cashed-up as everybody thought, but he had all the front in the world. He managed to negotiate the deposit down to 5 per cent – just $24 million – and was given three months to complete.

The only problem? Tinkler didn't have $24 million. He had to confess to his gobsmacked advisers he could only come up with $9 million in cash. He had plenty of assets, mind you, but it would be a major embarrassment to start selling them off so soon. Once again, Tinkler and his crew had to scramble to raise the money for something he'd already agreed to buy.

For a second time, it was Ken Talbot who came to Tinkler's aid. Tinkler explained what he wanted to buy and asked Talbot for a $15-million loan – and fast – to make up the deposit. Talbot put Dennis Wood on to Tom Todd, and within 48 hours a two-page agreement was drawn up. It was signed on 3 November 2009, the day before Rio was expecting its deposit.

Talbot Group Investments would lend Aston Resources $15 million for three months, interest-free. There were a couple of wrinkles: Talbot had a month to choose whether to convert part or all of the loan to shares in Aston, and would also be given the Indian and Sri Lankan marketing rights to the coal from Maules Creek. The whole deal was done so fast that it was full of holes.

Tinkler was alive to the danger: Talbot was putting down more money than he was, and so arguably had a better claim to the project. But Tinkler took the money anyway and paid Rio its deposit, leaving Aston only three months to find almost half a billion dollars, or else lose Maules Creek to Talbot.

Rio haughtily told the market that Coal & Allied would book an after-tax profit of $340 million on the sale. Its managing director, Bill Champion, said the Maules Creek project 'did not form part of [our] short-term or medium-term development plans'.

Tinkler told the *AFR* that Rio's senior executives, whom he had flown to London to see, had been 'absolutely astounded' with the price he was prepared to pay for Maules Creek, a project that was 'not in their 20-year plan'. They dismissed him as an idiot who had got lucky with Middlemount. 'Lucky idiot,' Tinkler recalled, 'but the lucky idiot got $1.2 billion for it six months later.'

*

Tinkler had his foot on Maules Creek. He still had to borrow the rest of the purchase price, which represented 10 times more than the fundraising for Middlemount, and in half the time. He needed help.

If Tinkler was going to take Aston public, the first step was to assemble a proper board. Tinkler was torn on whether to continue as a director of the company himself. On the one hand, he wanted to stay in control. On the other hand, all the feedback he got was that it would not help a float to have a major shareholder sitting on the board and, since he could be volatile and impatient about process, it would not suit his personality type. Christensen would be Tinkler's man on the Aston board.

Emulating Talbot, who had a former politician chairing Macarthur Coal, Tinkler invited Mark Vaile to join as chairman of Aston in mid-November of 2009. Vaile, who lived in Port Macquarie, had held the federal seat of Lyne, covering New South Wales' mid-north coast, for the National Party for 25 years until 2008. Vaile knew Buildev founder David Sharpe, who was close to the Liberal member for the federal seat of Paterson, Bob Baldwin, and may have provided the connection.

Tinkler spoke highly of Vaile as the representative of a rural electorate in which he'd spent much of his life. 'I'm a country boy,' he later told *Forbes*, 'there's no doubt about that, and Mark is a person I have always respected. He was a stock and station agent, which is what I wanted to start out as, but it didn't work out that way.' Tinkler also asked the former Sedgman chief executive Peter Hay, a veteran of BHP and Macarthur Coal recommended by Christensen, to join as a non-executive director.

Most importantly, if he was to pull off another Middlemount, Tinkler needed someone to handle the operational side of the business; Hamish Collins, who by then was running Aston Resources, was concentrating on the Queensland copper projects. Tinkler called Todd Hannigan at Xstrata Coal, whom he'd met while trying to sell his stake in Macarthur. Hannigan had been part of the due diligence team that had valued Macarthur at just $10 a share. When Tinkler got twice that much from ArcelorMittal, Hannigan had called to congratulate him. He hadn't heard back for six months. Now Tinkler invited him to join him at the Melbourne Cup, to watch the races and have a few beers.

At Flemington on Cup Day, Tinkler told Hannigan he'd enjoyed jousting with him. Hannigan was professional, he said, and had 'taken a bit of skin off us'. Would he like to join Aston? Hannigan came from Broken Hill and had mining in his blood. Maybe, he replied, if Aston had lined up a good asset.

Later in 2009, as the Maules Creek deal was firming, Tinkler renewed his overtures, and again Hannigan said he might be interested, if Tinkler could get the funding in place. In December, after paying Rio its deposit, Tinkler offered Hannigan a job as chief operating officer and 5 per cent of Aston. Tinkler could turn on the charm when he needed to, and his feisty attitude was infectious. As he told the *AFR*: 'The guys I have got at Aston would have spent the next ten years at Xstrata or Rio going absolutely fucking stale waiting for the guys in front of them to die or get their fucking commemorative pen before they get a chance to move up the fucking ladder.' After meeting with the Aston board in Newcastle, Hannigan jumped.

For Tinkler, this was to prove a crucial appointment. The advisory work on Aston's fundraising was given to the tiny merchant bank connected to Vaile, BKK Partners, which for a time also employed former federal treasurer Peter Costello. BKK drew up a list of possible Asian financiers for the Maules Creek acquisition. After looking this over, Hannigan suggested adding a name: Raymond Zage, a director of Noonday Asset Management, the Singapore arm of a big US hedge fund, Farallon Capital. Hannigan had dealt with Zage when he was at Xstrata Coal, evaluating opportunities in Indonesia.

Earlier in the decade, Zage had helped finance Indonesia's powerful Bakrie family as it bought coal mines in Kalimantan from the likes of BHP and Rio for a song. Both were caught napping by Indonesian laws that required a progressive sell-down to local investors, and had to stage rapid exits. Zage had watched in amazement as the Bakrie Group had managed to both cut costs and double production at mines such as the giant Kaltim Prima Coal. Zage had realised then that the major mining companies like BHP and Rio sometimes got it badly wrong.

Hannigan now believed that Zage might like to hear that Aston was buying Maules Creek from Rio. So the Aston team polished its pitch, and BKK organised two days of meetings with private-equity and hedge-fund investors in Singapore and Hong Kong.

When he joined, Hannigan had been told by Tinkler that Aston could contribute up to $100 million to fund the purchase of Maules Creek. In mid-December, just as Hannigan was about to begin his overseas roadshow, Tom Todd fessed up: $15 million of the deposit had actually been borrowed, and there was no more money to be had. They would have to raise fully 95 per cent of the purchase price of $480 million. Hannigan felt duped: would he end up begging for his old job back at Xstrata?

In January 2010, Tinkler, Hannigan and Todd took off for Singapore in the private jet. Hannigan's excitement was coloured by the prospect that the whole Maules Creek deal was threatening to unravel. The actual pitching was left to Hannigan and Todd – the duo was soon known as 'the two Todds'. Tinkler, heavily overweight, was aware he was not very presentable and preferred to spend the day with his personal assistant, Sally Reynolds; he'd get a briefing from the two Todds each evening over dinner.

Tinkler believed Aston would get its funding from Noble, and was working on Will Randall personally. But Noble was still smarting from the Middlemount deal, and was particularly pissed at how quickly Tinkler had taken his money off the table. Now they were after revenge, trying to screw Aston down on every possible angle: squeezing a higher marketing fee, grabbing marketing rights for the life of the mine on all coal, taking a larger stake in the company and diluting Aston.

In truth, while he should have been focused on raising $500 million, only six weeks out from settlement Tinkler was considering whether to take out a three-year lease on new offices for Aston in Brisbane. It was crazy: Aston didn't even know yet if it had a business, but in Tinkler's mind the deal was done – he had already moved on. The lease was signed and Sally Reynolds got the builders in: a whole side wing was to be created, with an enormous office for Tinkler. By the time the renovation began, however, Tinkler had already changed his mind, having decided instead to buy a new head office for the Tinkler Group, and in June 2010 he forked out $8 million for a two-storey building at 366 Queen Street.

Most of the investors the two Todds met were dubious, both about whom they were hearing from and about the high price Aston had paid for Maules Creek. The one important exception was Noonday: Raymond Zage was immediately hooked. He liked the asset. He liked the fact that Rio was the vendor. With coal prices having started to recover in January 2010, was the Maules Creek deal in the money already?

Hannigan and Todd had landed a big fish. Zage had connections with some immensely wealthy investors, such as the Wilmar International chairman and CEO Khoon Hong Kuok – thought to be the most likely successor to his uncle Robert, known as 'Mr Asia' – and his deputy, Martua Sitorus. Both were key investors in Noonday's funds; in special situations, Zage would offer them opportunities to invest alongside him, on their own account. As Hannigan and Todd were heading home in Tinkler's jet, Zage emailed them with indicative terms to lend Aston half the Maules Creek purchase price, roughly US$250 million. They were halfway there. Tinkler was still assuming that Noble would fund the other half.

A period of due diligence followed, and Zage flew from Newcastle to Narrabri, over the port, rail and project site. Behind the scenes, Zage was talking to Randall, and he sensed that Noble was getting cold feet. He sounded out Kuok and Sitorus about funding the whole Maules Creek deal, not half of it, so Kuok and Sitorus then made their own trip to Newcastle.

Tinkler, too, felt the talks with Noble were failing. Sources close to the deal were quoted in the *Sydney Morning Herald*, claiming that Tinkler 'realised they were going to squeeze his balls and chuck him out of it'. On 4 February 2011, just days out from settlement, Tinkler, Todd and BKK's John Anderson went to Singapore to meet with Randall. As they waited in the grand Fullerton Hotel, Randall emailed to tell them he was caught up over a bond issue and asked them to 'sit tight'. Tinkler snapped:

Hi Will

Have tried many times to contact you ... Noble's attitude to this transaction has been disappointing for the quality of opportunity this is and we do not have any more time to waste whilst you and your team continue to negotiate by yourself. We introduced Noble in good faith to Noonday and since this time Noble has done nothing but try to side with them to squeeze Aston. Problem with this is that Noonday are interested in building the value of Aston the entity they are investing in, not partnering with Noble to transfer value out of Aston. Aston can close this aquisition so that is what we will go and do.

I think Noble is a very different organisation to the one I remember and if you can find guys like my team who can pull together opportunities like this then good luck to you, I wish you the best with them. We are shaking in our boots at the thought of competing with your management team in Australia.

Our whole discussions have been a waste of our time and you can not even show me the courtesy of returning my call or meeting me after I have come to Singapore to meet you. This is the third occasion it has happened on this deal so 10 days out we can put up with this shit no longer.

Good luck with your future investments in coal, one day when your head comes out of your arse you might remember who led you into the best of them.

Regards
Nathan

It was a classic spray. From this point on, discussions would focus on Farallon funding the whole Maules Creek transaction. Meanwhile, Zage's bosses at Farallon's head office in San Francisco wanted to feel comfortable about lending up to half a billion dollars to Tinkler. With more than US$26 billion under management, Farallon was one of the most storied hedge funds in the United States. Its founder, Tom Steyers, a multibillionaire who then drove an old Honda Accord and has since quit Farallon to campaign with Michael Bloomberg on climate change, had pioneered an 'event-driven' style of investing, patiently looking to buy distressed or mispriced assets.

Farallon had made a fortune in Indonesia, where Zage was its point man, but it had nobody directly employed in Australia. Steyers called an Australian mate, Mark Carnegie, a fellow former merchant banker he knew from his time in New York. Carnegie turned to his own partner, John Singleton, who vouched for Tinkler. Farallon was over the line. For their efforts, both Singleton and Carnegie got in on the ground floor at Aston, picking up small stakes that would soon be worth millions.

One problem remained: everybody believed that Aston had put up the whole deposit itself. Only a handful of insiders knew the truth, and Tinkler had promised them he would somehow quietly find the money to repay Talbot, in order to get the whole thing off the books. But by February 2010, as the three-month deadline for settlement with Rio Tinto got nearer and nearer, Tinkler became increasingly narky. When asked about progress with the Talbot loan, he snapped back: 'Fuck off, leave that to me – you worry about Farallon.'

Whitehaven Coal, meanwhile, got wind that Tinkler was in trouble with his funding, and quietly told Rio Tinto that if Aston failed to complete for any reason, it would pay Tinkler's price. Realising it had missed an opportunity, Whitehaven now began running an aggressive interference strategy, dangling a $700-million bid for Maules Creek in front of Aston, but at the same time talking the project down in the market.

Aston pressed on. A settlement meeting was set up for seven p.m. on the evening of 16 February at Freehills' Brisbane offices. The timing was critical: settlement had to take place by midnight, which would be complicated by the time zones involved. Money in two tranches was coming from accounts on both sides of the United States, and had to clear in Rio's accounts in London within 24 hours. Farallon was there. Vaile was there. The wives of Hannigan, Vaile and Hay were there too, since many of the assets were in their names. The lawyers were there, and the documents were all laid out. But where was Nathan?

Literally minutes before the deal was due to be signed, Tinkler sent Hannigan and Todd a text message: *Meet me downstairs*. He was across the road at the Pig'n'Whistle pub, wearing shorts and thongs. He had a confession to make: he had not been able to come up with the money to refinance the Talbot loan. Despite everything Tinkler had riding on the Maules Creek deal, and despite all the time and effort that had gone into it, the young rich-lister could not scrape together $15 million from anywhere. The upshot was painfully clear: Rio and Farallon would have to be told.

Normally, hiding anything so significant until the very last minute would shatter trust on all sides of a corporate transaction. Convinced that the deal had been scuttled, Hannigan and Todd trundled back upstairs and told the assembled parties: 'We have a significant disclosure to make: there's an additional $15-million liability we haven't told you about.'

Farallon's man at the Brisbane meeting wanted to abort the deal immediately, but Zage, on the line from Singapore, kept his cool. (His pulse, apparently, never goes above 80 beats per minute.) After consulting with Kuoks and Sitorus, Zage confirmed that Farallon would stick with the deal, on the condition that Tinkler provided a personal guarantee over all his assets, and cleared the Talbot loan within 30 days.

Tinkler by now was storming in and out of the meeting, and hurling abuse at all and sundry. At one stage he literally walked out – he wouldn't cop the guarantee. Tom Todd had to go downstairs, sit with him on Queen Street, and coax him back into the meeting. It was two a.m. before a deal was reached with Farallon to take up the Talbot loan. It was four a.m. by the time the paperwork was signed. Nobody got much sleep, if any.

By nine a.m. the next morning, only $200 million had been deposited in Rio's accounts; the Kuoks and Sitorus were so keen to get on board that they'd deposited their money early. Rio met with

Aston and told its representatives that half the money wasn't there. They wanted to treat it as a non-refundable deposit, and were threatening to pull out of the deal altogether. Philip Christensen pointed out that nowhere in the contracts did it say time was of the essence. Rio would have to hang on.

At last the funds were cleared. Aston, hocked to the eyeballs to Farallon, was now the proud owner of Maules Creek. But Tinkler, who was embarrassed about the Talbot loan, and unhappy about the personal guarantee he'd been forced to provide, was in a deep hole. Ken Talbot, who was still waiting for his $15 million to be paid back, now decided that he might want in on the Maules Creek deal after all – even though in January he had indicated that he wouldn't convert his loan to equity. More than anything, Tinkler did not want to share the deal's potential upside with Talbot. The loan clause that gave Talbot the coal marketing rights in India and Sri Lanka had turned problematic, too; Todd Hannigan, who was now answerable to Noonday as well as to his employer, opposed Tinkler trading away rights that belonged to all Aston shareholders.

The Talbot loan had become a massive headache. Never keen to part with cash, Tinkler came up with another idea: would Talbot take a stake in Aston Copper instead? The company held interests in a remote, low-grade copper/zinc deposit near Mount Isa. Talbot mulled it over but wasn't interested.

More farfetched proposals followed; Tinkler was stretching the friendship. Talbot was hardly broke – he had no urgent need for the money – but a cynic might have said Tinkler was trying to avoid repaying the loan. Tinkler now had some ready cash, since in May he had offloaded 1.1 million of his unlisted Aston shares to Kuok for what was believed to be about $25 million – a deal that effectively valued Maules Creek at $600 million.

Then, on 19 June 2010, everything changed. Ken Talbot and the entire board of the explorer Sundance Resources were lost in a chartered plane crash in the Congo, where they were inspecting an iron ore project. At the age of 59, and with a fortune worth $965 million, Talbot was dead. Tributes poured in from the state's business leaders and politicians; Talbot's looming trial on corruption charges was all but forgotten.

In the aftermath of the plane crash, the Talbot Group had to drastically overhaul its investment strategy, since the six beneficiaries of Talbot's estate had a completely different appetite for risk to the company's late founder. The strategy immediately shifted from expansion to contraction, and the chair of the Talbot Group, Don Nissen, a former Commonwealth Bank executive, had to review investments in all 35 of the companies it owned or part-owned. The time for considering Tinkler's ambitious mining schemes had passed. Talbot's family wanted their money – for real this time.

Publicly, Tinkler eulogised Talbot, telling newspapers he would be 'irreplaceable as a mentor and friend to those who were fortunate enough to know him personally and … sorely missed by a very wide group of people across Australia'. But to the disgust of many around him, Tinkler continued to stall on the loan. The Talbot estate called in lawyers and threatened to sue, until Aston director Peter Hay intervened to urge sanity. Surely Tinkler would repay the widow of his late mentor? Tinkler sold the Kuoks a few extra Aston shares and the loan was finally repaid, but the episode left a foul aftertaste. What kind of person was Nathan?

Those close to Talbot still bristle at comparisons between him and Tinkler – or their supposed mentor–protégé relationship. Talbot, they say, unlike Tinkler, built and operated mines, rather than buying and flipping undeveloped tenements. Talbot, unlike Tinkler, was loyal to his friends and colleagues, and generous to a fault. Talbot, unlike Tinkler, paid his bills.

*

If 2009 was a good year to be buying assets, by 2010 it was not yet time to be selling them. But Tinkler's personal financial situation meant he needed to make a quick buck by flipping Maules Creek, in order to get himself out from under the yoke of the personal guarantee he'd just given Farallon.

The structure of the Farallon loans also dictated a quick sale. When the Maules deal settled, Farallon provided two main facilities: a 90-day loan of US$120 million, at up to 17 or 18 per cent – credit card interest rates – plus a three-year US$336-million loan, at a slightly more tolerable 12.5 per cent for the first year, but kicking up to 15 per cent thereafter. The total funding of US$456 million – which equated to $507 million at the prevailing exchange rate of about 90 cents to the US dollar – was enough to cover the balance of the purchase price and the $25 million owing in stamp duty, and to tide Aston over until a refinancing. The terms of the loans put huge pressure on Aston: any default would result in an explosion of warrants and, under an option agreement, give Farallon the chance to convert the debt to equity and take control of the company.

Aston had to move fast. There was no time for a thorough drilling campaign to prove up the asset, or for advance planning approvals. It decided on a two-track process, exploring a sell-down to a strategic partner and a public float at the same time. Xstrata was interested, but negotiations got bogged down over price. One morning Tinkler shocked everyone by turning up to work in a suit and tie, having decided it was 'signing day'. Tinkler and the Aston team were off to meet Xstrata's global coal chief, Peter Freyberg. He asked Tinkler to sweeten the deal by including marketing rights to the coal. 'If I make this deal any sweeter,' Tinkler poked at Freyberg, 'you'll get diabetes!' Everyone laughed, but the deal went nowhere.

Listing Aston on the stock exchange was now definitely looking like the best course of action. In a post-GFC market that was still offering slim pickings to the investment banks, there was plenty of competition to manage the float of Maules Creek. Goldman Sachs and Macquarie Bank were selected. But 90 days wasn't enough time to finalise the float. Aston paid through the nose for a third Farallon facility – this time worth US$132 million – to refinance the first. Credit Suisse, which had pitched unsuccessfully for a role in the float, was told that it could join in if it provided a backstop: a one-year facility of US$140 million, to be used only if Aston's initial public offering failed.

Aston had some breathing space but the environment was getting worse, not better. In May, Prime Minister Kevin Rudd announced that his government would introduce a new mining tax, the Resource Super-Profits Tax, provoking one of the most bitter fights between the government and the mining industry the country has ever seen. In the campaign that followed, mining companies big and small decried the uncertainty created by the proposed tax, and warned of the damage it would do to investment, particularly in greenfields development.

By June Rudd had been ousted, and the new Labor leader and prime minister, Julia Gillard, proposed a watered-down tax, the Minerals Resource Rent Tax. It had a lower rate, applied only to iron ore and coal, and extended generous credits for up-front capital investment. Most analysts thought the new tax would raise very little revenue. Still, Tinkler blamed the mining tax for putting international investors off the Aston float, saying in August that it had 'placed a huge question over sovereign risk in Australia'.

On the other hand, if there was one thing Asia still wanted, it was coal – particularly as stimulus spending kicked in across China, Korea and Japan. Coal prices started to rebound from their post-GFC lows: the industry experts' report prepared for the Aston prospectus noted that prices had doubled in recent Japanese contracts for semi-soft coking coal and PCI coal, to around US$170 a tonne, while thermal coal was back to almost US$100 a tonne. The market recovery was driving a renewed flurry of takeover activity.

Maules Creek was an attractive deposit. First, the all-important 'strip ratio' – of overburden to mineable coal, a measure of the amount of digging you have to do – was low, at 6.4:1. This meant that Maules Creek coal would be cheap to extract. Production costs were estimated at $56 a tonne, putting the project in a good position to compete with others for investment.

Second, Maules Creek was big – the seventh-biggest coal asset in Australia, according to one of the charts in the Aston prospectus, based on the maiden JORC-compliant reserve estimate of 356 tonnes, produced by the technical consultant Minarco.

There was a bit of history to this. In December 2003 Coal & Allied's accounts showed Maules Creek having JORC-compliant recoverable reserves of 160 million tonnes. But when the JORC code was updated in 2004, Rio's geologists took a conservative stance and downgraded the deposit, writing its reserves down to zero, while calculating that it had an economically recoverable resource of 680 million tonnes. At the end of 2008 Rio downgraded the Maules Creek resource again, to 398 million tonnes, after reviewing the economic assumptions it had used.

These variations had all been desktop studies – there had been no new exploration at Maules Creek since 1996. Aston, chucking Rio's conservatism out the window, hired Minarco to give it its JORC-compliant reserves figure, and lifted the resource estimate back up to 610 million tonnes – all without drilling a single new hole. Had Rio appointed an investment bank to sell Maules Creek, rather than doing the job itself, it might have done the same thing, and could well have generated a lot more interest in the asset.

Thirdly, the quality of Maules Creek coal was good, with a higher than average 'calorific value', or energy content, which should allow its thermal coal product to sell at a premium compared to the Newcastle benchmark price, and low ash or non-combustible content. Maules Creek also had a large proportion of semi-soft coking coal, which typically sells at a large premium to thermal coal.

Maules Creek was undoubtedly valuable, but how much was it worth? Aston hoped a float would value the project at $2 billion. After sounding out major investors in Australia and overseas, those expectations were lowered to $1.8 billion. Investors generally had one of two attitudes: some looked backwards, at how much Aston had paid for Maules Creek, and wondered why they were being asked to pay so much more for the project so soon, when nothing had changed; others looked forwards, at how much money could be made if the mine was developed as Aston was proposing. The first group of investors – particularly in Europe and Asia – were less likely to back the offering. The second group – predominantly in the United States – were more open to it.

Helping to placate sceptical investors were Aston's deals with Noble and the Japanese commodities trading house Itochu, which each agreed in advance to take a significant parcel of shares in the float, at the price they were being offered to the public. In July Noble agreed to pay $74 million for a 5 per cent stake – and received an offtake agreement giving it the rights to sell 1.5 million tonnes of Maules Creek coal each year – and in August Itochu agreed to pay $33 million for a 3 per cent stake, winning an exclusive period of four months to decide whether to take a strategic stake in the project.

Even with such significant backing, the float was touch and go. In July the failure of the $1.3-billion float of the major construction company Valemus cast a pall over an already depressed IPO market. Despite months of groundwork, roadshows and titillating items promoting the float in the media, Aston's first attempt – to raise $400 million at $8.20 a share – failed on 4 August 2010.

That day, Tinkler gathered the Aston board at his Martin Place office. Hannigan and Todd were spent; they didn't know what to do next. Tinkler, too, must have been on edge, since if the float failed, he was all but broke. The investment banks – BKK, Goldman Sachs and Macquarie – were nowhere to be seen. It felt like a war room.

Tinkler first told everyone to stop whimpering, then he put the domestic order book up on a whiteboard. Colonial First State had placed the biggest stock order, for $25 million worth; the rest was rats and mice. Full of bluster, Tinkler said he could make 10 phone calls himself and raise $10 million if he needed to; the truth was that none of his horsey mates had so far come through. Whatever else Tinkler said, it was sufficiently inspiring that everyone left the meeting feeling that the fundraising was feasible. When the time had come, Tinkler had shown balls. Leadership, even.

The next day, the chastened bankers came into Tinkler's office together and said they could get the float away if Maules was repriced at $1.2 billion. Tinkler thought about it briefly, then told them to go for it. It soon became clear that the banks had sounded out the key financial institutions already, but even at a cut-down offer price, the Aston IPO was a close-run thing.

At six a.m. on the day of the float, Goldman Sachs was nervous. It was déjà vu: the international bookbuild had raised $375 million overnight, but with three hours to go until the offer closed, the brokers were still short.

Tinkler texted Hannigan and Todd to meet him for coffee at the Westin Hotel. He was relaxed and joking: don't worry, it'll get done! Across the road at Tinkler's offices, Macquarie was briefing the rest of the Aston board. Just as it looked like the float was about to fall over – one word to the market about the shortfall and everyone would start cancelling orders again – a $100-million order came through from a major US fund manager. The float was away; in fact, it would end up being modestly oversubscribed.

Aston had done it, raising about $420 million from the issue of 67 million shares at $5.96 each, leaving $379 million in net proceeds, after the bankers, lawyers and other consultants had soaked up roughly $22 million in fees – and a whopping $19-million interest bill to Farallon, racked up between February and August, was converted to equity.

Done within nine months, in a decidedly tough market, the IPO of Aston Resources was a remarkable coup: the biggest coal float Australia had ever seen. When it was wholly owned by Tinkler, Aston had paid $480 million for Maules Creek. By selling a third of Aston for roughly $400 million in the float, Tinkler had effectively revalued Maules Creek at $1.2 billion. His own stake in Aston – 72 million shares, or 35 per cent – was worth $427 million at the float price. That was 18 times the deposit of $24 million he'd put down on Maules Creek just nine months earlier, or an annualised return of more than 2300 per cent! Lightning had struck twice for Tinkler.

Tinkler hadn't yet taken a big lick of cash out of Maules Creek, as he had with Middlemount. His stake in Aston was held in escrow for six months after the float, so he couldn't sell his shares immediately, except for estate-planning purposes (a wide loophole). But Tinkler again had a large stake in a publicly listed company, and the escrow agreement allowed him to pledge his shares to borrow up to $100 million. Tinkler was liquid again, and free of the loathed personal guarantee he'd given Farallon. For the first time in months, he could sleep easy.

In September 2010 Tinkler recovered his place at the top of *BRW*'s Young Rich list, with a fortune of $610 million. By the time the Rich 200 list came out in May the next year, a 64 per cent rise in Aston's share price had boosted his wealth just over the billion-dollar mark; he was only 35 years old. The magazine, estimating his fortune at $1.01 billion, congratulated him on the remarkable feat: 'He's done it: Nathan Tinkler has become a billionaire.'

The success of the Aston float spread a bit of wealth around: the shares issued to Hannigan and Todd were now worth $26 million to each of them at the float price, and Vaile and Hay now had stakes worth $10 million each. For Vaile, who was used to a politician's wage, it was a payday beyond his wildest dreams.

The faith of Ray Zage and his backers was also rewarded. Warrants issued at the time of the February loan had given them almost a quarter of the company: Farallon emerged with 8 per cent, worth $96 million, while Kuok and his family held 9 per cent, worth $128 million, and Sitorus held 3 per cent, worth $38 million. The stakes of Singleton and Carnegie, who now had a tacit watching brief over Aston, were worth $5 million each.

In round figures, 60 per cent of the $380 million float proceeds to Aston – some $230 million – went straight to Farallon, leaving Aston with net debt of roughly $150 million on its main three-year loan. The rising Australian dollar would help reduce Aston's debts, which were denominated in US dollars. The Credit Suisse backstop facility was never used.

Aston traded poorly at first. With only $150 million funding available for the development of Maules Creek, which was expected to cost $463 million, Aston still needed more money – a prospect which hadn't helped the float and weighed on its share price. In December, though, came confirmation that Itochu had paid $345 million for 15 per cent of the project, implying a Maules Creek valuation of $2.3 billion. By the end of 2010 Aston had the world in front of it. Its shares hit $10, it could fund its development, and it would soon refinance its expensive Farallon debt on more commercial terms.

Instead of celebrating the Aston float, Tinkler became increasingly jealous. In December 2010 a 'tombstone' trophy was made, commemorating the role of the various parties – Aston itself, the lender Noonday, investment bankers Goldman Sachs, Macquarie and Credit Suisse, lawyers Freehills and auditors Ernst & Young – in the success.

Such 'tombstones' – originally named after the sombre, small advertisements in the financial press – are a bit of puffery, mementos of a deal, and usually identify the businesses involved, not individuals. Tinkler, however, when he received his 'tombstone' (which he referred to as a 'milestone'), fired off a churlish email to Hannigan, Todd, counsel Melissa Swain and assistant Jodie van Gilst – making sure to copy in boardmembers Vaile, Christensen and Hay – with the subject header 'You Guys Are Great!':

Hi

Thanks for sending down the milestone of the Aston deal. How thoughtful to include me. I cant imagine why you would send me one when my efforts did not even warrant being ON the milestone as opposed to every other deadbeat that worked on it for 5 mins?

i am still around, after hand picking every single one of you for your job, still the major shareholder and still know the best way forward. I am sorry I am not in the office to keep everyones ego in check these days and stop heads disappearing up arses.

I wont let Aston become just another coal company! I will make changes before I sell shares and I will be there on Monday to start discussing them! I cant believe how quickly the culture of the place has gone to shit.

Enjoy your weekend.

Nathan Tinkler

Over Christmas he invited the whole board to 'Noorinya' for lunch. Private jets flew in to Newcastle from Sydney and Brisbane. At the gathering, Tinkler apologised for his behaviour but said he'd decided to sell out of Aston when his escrow was lifted. The mood was pleasant but restrained.

When the summer edition *AFR* wrote up the Aston float as one of the best deals of 2010, and proclaimed Tinkler a genius, he cheered up and changed his mind. It would be onwards and upwards.

*

Right after the float, Tinkler had celebrated by opening up to the *AFR* in one of his periodic self-serving interviews. The story kicked off with quotes from Tinkler that addressed the perception he was an asset-flipper, rather than a genuine miner. 'If I was just a wheeler and dealer,' he said, 'I might have put my cue in the rack and walked away rather than trying to build a portfolio of operating mines. I think over the next decade people will start to see me as a developer.'

Tinkler also mused on his good fortune at snaring Maules Creek for $480 million, observing that 'the world has probably pulled itself out of the rut a lot quicker than we expected when we did pony up with the money. What gave us confidence to do that was that coal markets and prices haven't really taken the battering that the credit world had taken.'

Takeover activity in Australian coal was still running hot and, surveying the landscape, Tinkler commented that American companies such as Peabody were likely to be more disciplined buyers than Indian groups such as Adani, which the previous week had paid $2 billion to buy a deposit in Queensland's undeveloped Galilee Basin. 'They seem to be buying assets around the world on the size of tonnes in the ground at the moment, rather than the quality and cost to excavate them,' he opined. 'They seem to me to be a bit of an immature investor and probably need to tackle a few projects to really get an understanding as to why the Asians that we're seeing in the market do it the way they do.'

Tinkler also reckoned that experienced Chinese players such as Shenhua, the world's largest coal supplier, would make their presence felt. In 2008 Shenhua had paid $300 million to the New South Wales government for the right to explore the Watermark project in the Gunnedah Basin. 'I don't think the likes of Shenhua are going to be happy just sitting at Gunnedah drilling some holes trying to prove up one asset,' Tinkler stated. 'They produce a lot of tonnes back home and they are very skilled in the underground mining arena. I would think that as Australian mining shifts more towards underground, that will become a lot more appealing to the Chinese.'

With his Patinack hat on, Tinkler also offered some loose comments on the conflict between coal mining and horseracing in the Hunter Valley, which had seen the state government block the $3.6-billion Bickham mine, near Scone. Tinkler said the government should avoid offering mining rights in parts of the valley where it would 'very obviously compete' with racehorse breeding and training:

> There's parts of the Hunter Valley which, you can't say they're off limits to mining ... but certainly areas like the Segenhoe region and some of those areas round where the major studs are, they need to be protected.
>
> But to say, in other areas, just to have a blanket rule which says there will be no more mining in the Scone or Muswellbrook shires, I think that's a bit one-sided. There are areas in both those shires that are not going to interfere with the horse industry, which is everyone's major concern.

Tinkler sounded like a winner, holding forth, but he was being altogether too cocky. He had great confidence in Aston – but what had the investors in it actually bought? Despite its compelling economics, the Maules Creek project would prove challenging, to say the least. Aston's prospectus was forecasting that all mine approvals would be given in 2011, with the first coal produced by May 2012 and the first sales by December. That would turn out to be wildly optimistic. Final approvals for Maules Creek would not be given until July 2013; at the time of writing, the mine is facing a legal challenge from the Environment Defender's Office, claiming the final federal approval was invalid because the minister allowed himself to be influenced by leaks from New South Wales resources minister Chris Hartcher's office.

Tensions over the collision of mining and agriculture could not be dismissed so easily. Knowingly or unknowingly, Aston had stepped into a bitter fight, both with local farmers who were concerned about the impact of more coal mines on their land, and with an anti-coal movement that was driven by concern about climate change. Opposition from farmers and environmentalists would prove a significant obstacle to mining generally, particularly as it expanded into food-bowl areas such as the Liverpool Plains and Queensland's Darling Downs.

Aston was by no means the only Gunnedah Basin miner caught up in the debate. BHP Billiton found itself facing a residents' blockade at its Caroona project, which attracted national attention and a Senate inquiry in 2009, and drew criticism from conservative politicians like Barnaby Joyce, Bill Heffernan and independent Tony Windsor. China's Shenhua had the same headaches at its nearby Watermark project; it was forced to spend millions of dollars buying up prime farmland around its proposed mine site.

Unlike Caroona or Watermark, as Aston's prospectus noted, Maules Creek was east of the main alluvial floodplain zone of the Namoi River, and so was 'remote from the fertile and high agricultural value Liverpool Plains'. There were no stud farms or vineyards nearby. In fact, the impact of the mine on neighbouring properties was minimised because most of the digging would be in the Leard State Forest. This was presented as a benefit, but Aston didn't bet on trenchant opposition from Phil Laird, a sixth-generation local farmer whose family the forest was named after.

Laird pushed for the mine to go underground. Rio had originally proposed a combination of open-cut and underground mining, which was approved by the state government in 1991. Laird formed the Maules Creek Community Council in 2010, to oppose both Aston's project and the expansion of Whitehaven's Tarrawonga and Boggabri joint venture with Idemitsu. The combined effect of the three projects would be to clear two-thirds of the forest, which was one of the largest remaining woodlands with the critically endangered white box gum on the Liverpool Plains, which also provided habitat for endangered species such as the swift parrot.

Aston's prospectus acknowledged the presence of the high-conservation-value timber, but noted that it had been cleared before and that land elsewhere could be bought to offset the damage its mine might cause. But the offset strategy itself would be expensive and controversial, with ecologists contracted by the Maules Creek Community Council and the Northern Inland Council for the Environment arguing that the purchase of two nearby farms, Wirradale and Mount Lindesay, was completely inappropriate and would not compensate for the loss of natural habitat for endangered species.

Then federal independent MP Tony Windsor toured the Maules Creek site and the land used for offsets in a helicopter with the then environment minister, Tony Burke, and came away dubious. 'I don't think you can just say "We'll remove these endangered species and replace it with some cleared farming country, which we won't do anything to, so eventually it will become the same as what it was 100 years ago and offset what we're removing,"' Windsor told the *Sydney Morning Herald*. 'I just have a bit of difficulty getting my head around that. I don't know all of their offsets but to the naked eye looking down on it you look at what they're wanting to remove and what they're wanting to replace it with – the country outside the forest – there (would be) very little timber left.'

Around this time, amid planning for the Maules Creek project, Tinkler appears to have made the only political donations of his life. Ahead of the March 2011 New South Wales election, when the electorate took a baseball bat to a corrupt Labor government, Tinkler personally gave $50,000 to the New South Wales National Party, whose member Kevin Anderson took the seat of Tamworth, which included Maules Creek. After taking legal advice, Todd Hannigan and Tom Todd gave $5000 each, as did Peter Hay. In May the Tinkler Group also donated $22,000 to the federal Liberal Party. It was all arranged by Aston's board member Mark Vaile, who also donated.

None of these donations were disclosed as part of the Maules Creek development application, which was made in the name of a subsidiary, Aston Coal 2, of which Tinkler was not a director. Two years later, after pressure from the New South Wales Greens, the state planning department launched legal action against Aston Coal 2 in the Land and Environment Court for failing to disclose the Hannigan and Todd donations. Greens MP Jeremy Buckingham told the ABC's *Four Corners*:

> You go to the Department of Planning website and there's a sworn statement that there has been no political donations made by these entities. You go to the Electoral Commission website and there they are, the people running these companies, the key figures making significant political donations. It completely undermines confidence in the system and the

community is left in the dark. Tinkler's manipulating the system so he can make a donation, and use this associated entity … to game the system, make the donation, get the application in and ultimately get an approval for a contentious mine … whilst hiding a fifty thousand dollar personal donation.

Ultimately, Whitehaven – which by then was the owner of Aston Coal 2 – pleaded guilty and copped a fine.

In the end, however, it was the new Coalition state government, rather than the federal Labor government, which held up Maules Creek the longest. Two years after Aston's application, in March 2012, the NSW Planning Assessment Commission's first review report flashed amber – proceed with caution – and the rest of the approval process was expected to be a formality. Instead, the approval became bogged down. The state planning minister, Brad Hazzard, admitted to a gathering of 400 people at Gunnedah Town Hall in the same month that it was 'illogical' to allow an open-cut mine within a state forest. Final approval by the NSW Planning Assessment Commission was not given until October 2012.

Federal approval then took another nine months, after the project got caught up in a political fight between resources minister Chris Hartcher and federal environment minister Burke. Lawyers from the Environmental Defenders Office NSW would later argue that the dispute invalidated the Commonwealth approval, and would mount a Federal Court challenge against it, which is still ongoing.

In short, everything that could go wrong in the planning process for Maules Creek did go wrong. The lost years eroded its profitability, particularly as the coal market continued to shift. In its prospectus, Aston's consultants stated that they did 'not anticipate a significant decline in benchmark thermal coal prices in the medium term'. That prediction bore true for 2011 but went wrong in 2012, when coal prices fell. Maules Creek's budget blew out from $463 million in its 2010 prospectus to $760 million by 2013.

Tinkler was dismissive of climate change – he wasn't going to be around in 100 years' time, so why should he worry? He did not realise that Maules Creek would become a prime target for climate activists. While not on the scale of the mega-mines being planned in Queensland's Galilee Basin, Maules Creek would produce 10 million tonnes of coal per annum for 30 years or more, and emit 30 million tonnes of CO_2 a year – roughly equivalent to 5 per cent of Australia's total emissions – if the emissions from burning the coal overseas were included.

In 2012 a young Newcastle student, Jonathan Moylan, representing a tiny group called Front Line Action on Coal, set up a protest camp within the Leard State Forest. He would soon catapult Maules Creek onto the national stage by issuing a fake ANZ press release, purporting to withdraw project finance on reputational grounds. It was a grave step: a wire service took the statement at face value and reported it. For 23 minutes, the market was misled: shares in Whitehaven (which by then owned the mine) plunged 9 per cent before a trading halt was called, and a handful of investors who sold lost some $450,000 combined. Moylan was prosecuted by ASIC, with a trial set for November 2013. He faces up to 10 years in jail.

In August 2010, however, Maules Creek's planning woes lay in the distant future. Tinkler was gung-ho, saying the mine would come into production and 'just churn cash like nothing else … in fact, if that asset is not making a billion dollars a year by the end of five or six years I will be fucking astounded.' The Aston float was a great deal for Tinkler, his executives and directors. It was also a great deal for Farallon and its wealthy backers, who not only got all their money back promptly – and with interest – but also got substantial stakes in Aston, which were soon worth hundreds of millions.

More ominously, Maules Creek marked the beginning of Tinkler's relationship with Farallon's Ray Zage, which opened up a new source of finance to him but on terms that were far more aggressive than anything offered by normal banks. For Tinkler, who had no fear of debt, this was dangerous. He would increasingly come to rely on finance from Farallon.

'I like risk, I like leverage and that's something that's served me well in the past and will serve me well in the future,' Tinkler told Bloomberg in a 2011 interview. 'I have an aggressive mindset in the way that I do things and the way I deliver opportunities.'

True to his word, Tinkler would allow his debts to balloon to monstrous proportions. The Aston float would prove to be his last chance to get out with a motza – to quit while he was ahead.

Instead, Tinkler celebrated by upgrading to a bigger private jet, one capable of getting him from Sydney to Singapore direct. The $16-million, triple-engine Dassault Falcon 900C had plush leather seats for 14 in a gold-plated interior featuring gaboon veneer cabinetry; there was also a galley with a wide oven and espresso machine, CD and DVD player with dual 18-inch monitors. Tinkler had committed to buy the jet from Paul Little, the former CEO of Toll Holdings (and later chairman of the Essendon Football Club).

The only thing was, with most of his wealth now tied up in his Aston stake, Tinkler had trouble finding the money to settle. He tried to pull out, but couldn't: Little wouldn't stand for it. Eventually, Tinkler got a $13-million loan from GE and was able to pay what he owed Little for the jet … at least, most of it.

6.

TINKLERTOWN
THE WHITE KNIGHT RIDES IN

On any list of Australia's quiet achievers, Newcastle must be near the top. The world's biggest coal port, it generates a third of New South Wales' exports, despite having only 8 per cent of the state's population. It is the largest regional economy in Australia.

While the working port is Newcastle's lifeblood, and the decline of manufacturing is gradual, the former 'Steel City' has had to fight to slowly stake out a new future, built around the university, the hospital, tourism and lifestyle – all without much help from politicians, federal or state. Its redevelopment has been slow and patchy, and the city's grungy industrial roots are never far from the surface. The boardwalk along the waterfront looks great, but two streets back, along the main drag of Hunter Street, block after block remains vacant and derelict. It's the seventh-biggest city in the land, but Newcastle has the pace of a big country town.

In the wake of the Aston Resources float, Nathan Tinkler was at the peak of his wealth and power, and his remarkable rags-to-riches tale – from tradie to mining mogul – was retold over and over again. Perhaps nowhere did it resonate more strongly than Newcastle, his adopted home town. 'There was a great deal of well-wishing in this neck of the woods,' says veteran journalist Neil Jameson, 'because he was a pit electrician from Muswellbrook who'd obviously worked hard as a young man, and gone from very humble beginnings to an absolute fortune. The subliminal message was quite clear: "Why couldn't that be me?" And all around the Hunter everyone said: "Good on him, good luck, Nathan."'

As a mogul who'd worked at the coal face, Tinkler was a novelty; his background in the pits won him a kind of instant cred with the well-schooled bankers and fundies of Sydney and Melbourne. Tinkler was not to be underestimated. He was a bona fide Newcastle coal baron, a tag he wore in press reports whether he liked it or not.

Unlike Australia's other billionaire mining moguls, Gina Rinehart, Andrew Forrest and Clive Palmer, who relished the national stage, Tinkler didn't make forays into politics and didn't foist his views on all and sundry, whether on climate change, the mining tax, industrial relations, foreign investment or anything else. Instead, he was passionate about Newcastle going places, becoming a city on the up and up.

Tinkler had lived and worked in the Hunter Valley for some 15 years, and he shared the region's preoccupations: coal, horseracing and footy. Although he had divided his time between Newcastle and Brisbane since 2004, he was genuinely *of* Newcastle: even when he was getting around in his Ferrari, its custom plates – 'PIT LECO' – were celebrated as evidence of his class loyalty. He was a blue-collar worker done good.

For a time, Tinkler would become a kind of benefactor Newcastle had never before seen. As local ABC Radio journalist Aaron Kearney explained in 2011, Newcastle and the Hunter region had working-class roots and were 'used to being treated poorly':

> There are all sorts of data about how the revenue that is generated by this region way exceeds anything that is returned to us by government. Infrastructure-wise we are the poor cousins of Sydney, and, you can mount an argument, even some regional areas. It doesn't get nice things, it's not used to nice

things, and there's an argument that it doesn't like nice things because it's never had them. In some ways Nathan Tinkler has come and played Santa Claus in this region in a way governments have never been able to do ... all of a sudden Dad won Lotto, one of our own won Lotto, and started buying nice things for the region. That's wonderful and we're still in that honeymoon period. The big question for the Hunter is how long will that honeymoon last, and what will be the circuit breaker in that? Quite what will Nathan Tinkler expect in return from the region for his generosity? If the answer's nothing, then Santa Claus has indeed come.

Tinkler did give back to the area. There is no ready tally of how much he contributed to local charities, but through one entity or another he certainly donated to the John Hunter Children's Hospital, Ronald McDonald House Newcastle and the Westpac Rescue Helicopter. On the other hand, the Tinkler Foundation – which Tinkler talked about for some years but only legally established in mid-2011 – never quite got up and running. While it has a CEO, it is still not registered as a charitable organisation.

Undoubtedly, a lot of Tinkler's giving went under the radar. One nearby resident remembers him helping out a family in Kendall, his dad's home town, when their house burned down. A business associate remembers Tinkler ducking out of a meeting to take a call in the wake of the devastating flooding in the Lockyer Valley in January 2011, which killed 13 people. Tinkler walked back in and said he had just given $50,000 to help the victims.

In sport, Tinkler would get behind Surfest, held at Merewether, the biggest surfing competition in Australia and Newcastle's only international sporting event. He sponsored the women's comp when it was about to go under, and worked at one point with local legend and former world surfing champion Mark Richards to bring Kelly Slater to the event. Tinkler was keen to bring back Newcastle's netball and basketball teams. He sponsored local motorbike rider Jeremy Summers. The list goes on. But none of that compared remotely to the impact he made when he took over Newcastle's two football teams.

*

In September 2010, soccer's governing body, the Football Federation of Australia (FFA), stripped the property developer Con Constantine of his licence to run the Newcastle United Jets, one of the more successful clubs in the new A-League national competition. The Jets had won their first premiership in 2008 but had then slumped to collect the wooden spoon the following year.

Founded by Constantine in 2000 out of the remnants of the Newcastle Breakers, the Jets quickly built up a loyal following, despite their haphazard performance. Constantine owned western Sydney's Parklea Markets and a couple of local papers in Newcastle. He was supposedly worth some $200 million but his business was teetering in the wake of the GFC.

Constantine had poured $15 million into the Jets over the years. In 2010 he had just signed a new 10-year licence agreement with the FFA when he hit the wall: the team could not meet its short- or long-term liabilities, it emerged, and there were rumours that Constantine was being chased by the ATO. The club had to be given an FFA lifeline so that it could pay wages to its players and staff. Constantine insisted his problems were short-term and would be solved by selling off assets, but after a month's negotiations the FFA decided it could not tolerate the risk to the club.

Soccer legend Craig Johnston, the Newcastle-born expat who played for Liverpool and was Australia's most decorated footballer, told SBS how important the team was to the city. For him, it was unthinkable that the club might fold:

Some fans, you see, might not understand how vital Newcastle is to Australian football and vice-versa and what rich heritage the whole region enjoys … The club has profound Geordie links thanks to the English miners who came here at the turn of the century and helped start up a football club. Football clubs in Newcastle are among the oldest in Australia, some even older than a few English Premier League clubs. The very name of the city, plus those of such suburbs as Stockton and Gateshead, are proof of Newcastle's strong links with the English city in the north-east. For people in the Hunter, coal mining and football are everything. How can we let this heritage die?

Behind the scenes, the FFA had engaged a consultant, the well-known sports administrator Ken Edwards, to ring around potential wealthy donors in Newcastle and try to bail the Jets out. For more than a decade Edwards had been the chief executive of Sydney's Stadium Australia, which had hosted the 2000 Olympics, the 2003 Rugby World Cup and the Socceroos' heart-stopping World Cup qualifier in 2005. But his had been a tough gig: despite holding a fond place in the nation's heart, the publicly listed stadium was a white elephant commercially, having faced constant financial travails until it was taken over by its lender, ANZ, in 2007. Edwards, who was widely respected, had resigned in mid-2009 at the age of 52, and was freelancing when the FFA got in touch.

Barely knowing who Tinkler was, Edwards cold-called the young tycoon. The reception was warm: 'How much do you need?' Edwards answered: '$2 million.' Tinkler replied that it would be a pity to lose a Newcastle sporting flagship for the sake of just $2 million; he was in.

Stoked, Edwards reported back to the FFA's president, Frank Lowy, and was surprised to be rebuffed. As Neil Jameson told *Four Corners* in 2013: 'Lowy said, "Look, you've got Australia's richest man under 40 on the line here, surely you can get more than $2 million out of him – it'd be great if we could actually sign him up to a 10-year contract and make him part of the football family.'" Edwards went back and tried.

Suddenly, Tinkler had become Newcastle's white knight, saving the town's struggling soccer team. It was all unplanned and it all happened very quickly. When the deal was announced on 22 September 2010, Tinkler was blunt: he had 'no desire to own a football club', he said, and he was not trying to make money out of it. Rather, as a proud Novocastrian, he knew the importance of the Jets to the city and to the A-League:

> My family and I have been supporters of the Newcastle Jets for years and we are delighted to be able to step in and support the club for the city of Newcastle [and] provide a nursery for the development of the game for the thousands of young football players in the Hunter region.

Holding his own press conference in Newcastle that afternoon, a tearful Constantine was clearly devastated, accusing Lowy – whose Westfield empire was an aggressive competitor in the retail property business – of going behind his back to negotiate with Tinkler. 'I can't tell you what took place but maybe Frank Lowy doesn't want me to be here,' he said. 'There was no reason for FFA to pull the pin unless they were dealing with Tinkler over the last few weeks.'

He was right about that: the Tinkler Sports Group had been created only on 3 September. In the end, that didn't matter: despite Constantine's assurances, his financial problems were not short-term, and the Parklea Markets went into receivership the following year. He faced a long struggle to regain control of his business, which he finally did in 2013.

The people of Newcastle were overwhelmingly grateful to Tinkler for rescuing the Jets. Tinkler's first move was to appoint

Edwards as executive chairman, and he quickly became the public face of the Tinkler Sports Group. Comfortable with the media, Edwards assured journalists that Tinkler wanted to put in place a proper governance structure at the club, including an advisory board of knowledgeable Newcastle people; he planned to 'run the Jets as a community asset':

> Nathan genuinely believes this community needs national teams and he's using his investment to buy time to identify whether Newcastle really does want to have this club. This is not an open chequebook. He wants it to be run properly and for Newcastle to show that it will get behind it.

Tinkler's initial deal was to last a season only, until the end of March 2011, and included an option to extend his ownership of the club. But Tinkler was delighted by the public response to the bail-out, and almost straight away a 10-year agreement was being hammered out. The FFA threw everything at it. Lowy himself duchessed Tinkler, inviting him down to Westfield's Sydney offices to sort out a deal face-to-face, mogul-to-mogul. Finally, Tinkler took the plunge, agreeing to pay a $3.5-million acquisition fee, plus a $1-million licence fee, staggered over time, to take over the Jets until 2020. He also agreed to kick in $1 million a year to the development of junior footballers in the Hunter. The whole deal was announced in early October.

It was all systems go. Edwards' strategy was to win hearts and minds: he would boost the club's profitability by dropping the price of membership and linking it to family-friendly ticketing: cheap seats for the fans, 11 games for $100. It was a hugely popular strategy that would see Jets' membership grow from 500 to more than 10,000 within 18 months.

A few weeks after taking over, Edwards unveiled a masterstroke: the Jets would host the LA Galaxy, with an appearance by the

American club's star striker David Beckham. Edwards told the *Newcastle Herald* he'd received a call from the FFA, asking if the Jets were interested; the cost was $2 million. 'I rang Nathan,' he said, 'and after a 30-second conversation he said "absolutely".' The biggest football star on the planet was coming to Newcastle, and as far as the media was concerned it was all thanks to one guy, Tinkler.

The precise extent of his contribution wasn't made clear, but it is certain that the New South Wales government also contributed a fair whack. The 'Becks' coup for the Hunter was announced by the premier, Kristina Keneally, who saw an opportunity to trumpet Australia's soccer credentials in the lead-up to Australia's failed bid to host the World Cup in 2022. 'We want to show the world we are ready,' she said.

To top it off, when the Galaxy arrived, the match was a sellout. The Jets prevailed 2–1; despite flashes of brilliance, Beckham failed to score. Tinkler, as usual, didn't turn up.

The truth was that he didn't even like soccer; he was a league man. Journalist Neil Jameson recalls that Tinkler 'didn't hide that fact one bit: he said he knew very little about soccer but … he was willing to learn'. Amid the general surge of goodwill, Tinkler and Edwards brought in some good people – including Jameson, who joined the Jets' community advisory board. Coach Branko Culina, who had helped the Jets qualify for the finals in the 2009–10 season, was re-signed for four more years.

The Jets fared badly in 2010–11, missing the finals, but Tinkler's outfit was focused on the future. In February, Culina's son Jason, a World Cup veteran for the Socceroos and onetime Australian captain, was hired as the Jets' marquee player on a three-year deal worth $2.6 million. He had just had knee surgery but was expected to be fine in time for the next season. The road ahead seemed smooth.

*

Owning a sports team can be both humbling and gratifying for someone in business. You're suddenly in the spotlight, with thousands of obsessive fans picking over the club's every move. The positive reaction Tinkler received after his Jets takeover spurred him to buy a team in the game he'd played and loved. But the Jets had been a basket case in need of a saviour; the Newcastle Knights would have to be fought for and won.

The chairman of the Knights in 2011, former player Robert Tew, had taken over in 2008, with the explicit objective of looking at ownership alternatives for the struggling club, including full or part privatisation. There are half-a-dozen privately owned teams in the National Rugby League, including Manly, the Brisbane Broncos and the Melbourne Storm. Around this time, the foundation club South Sydney was beginning to prosper under the ownership of actor Russell Crowe and businessman Peter Holmes à Court; in 2006 they'd paid $3 million for a 75 per cent stake in the club, in a deal the Rabbitohs' members had approved by the barest margin, a mere 32 votes. Since then, the Rabbitohs' membership was growing, revenues were rising and the team was starting to win again.

When he was considering buying the Knights, Tinkler went to see Crowe and asked him if it was worth it.

Crowe: How many kids have you got?

Tinkler: Four.

Crowe: Well, do you need 40 more? You are held accountable for every fuck-up these guys make.

The story of rugby league in Newcastle is a study in satellite city pride and envy – and in neglect by the code's bigwigs in Sydney. Newcastle had provided one of the foundation clubs for the original Sydney league but had withdrawn in 1909. Although ensuing decades had seen a constant supply of champions from the Hunter Valley to the New South Wales Rugby League (NSWRL) competition and beyond – including greats like Clive

Churchill – the Newcastle Rugby League was treated as a fiefdom, kept apart from Sydney. So when, after the demise of the Newtown Jets, the NSWRL invited Illawarra and Newcastle to join the competition in 1982, the Newcastle Rugby League declined; Canberra joined instead.

The threat of competition from Australian rules football and the new Sydney Swans forced the NSWRL to expand again in 1987, and it took in new teams from Brisbane, the Gold Coast and Newcastle. As Neil Jameson records in his book *Our Town, Our Team*, when the Knights took the field at home for the first time in a pre-season match against then premiers Manly in 1988 – and *beat* them 22–12 – Newcastle went nuts. The longing that had built up over decades to see a homegrown Newcastle club compete with the best Sydney and the rest of the country had to offer was unleashed.

The Knights' place in history was cemented 10 years later, when they won their first Grand Final in a fairytale 22–16 victory, again over Manly; the winning try came in the last six seconds, scored by Darren Albert off a typically brilliant inside pass from Andrew Johns, the Cessnock-born rugby league immortal. That victory was even sweeter as it came in the middle of the bitter Super League war, in which the Knights stayed loyal to the ARL, while a rival club, the Hunter Mariners, played in the Murdoch-owned competition.

That unforgettable 1997 win ushered in a purple patch for the Knights and Johns, who went on to captain Newcastle – leading his team to another Grand Final win in 2001 – as well as the New South Wales and Australian representative sides. He ultimately became perhaps the most awarded player the game has ever seen.

Despite their place in Newcastle's heart, the Knights had always struggled financially, mainly because they lacked a licensed club – and therefore a stream of poker machine income – to subsidise the team. The year's bills were always paid for by the sale of next year's season tickets, in a kind of serial insolvency. The team never had a proper place to train, or decent player facilities. The International Sports Centre Stadium at Turton Road spent years under refurbishment, first with the construction of a new eastern stand – later renamed after Andrew Johns – and then with a new western stand. The construction itself hurt ticket sales, and the club wound up in a long-running legal dispute with the New South Wales government.

Robert Tew, a valuer by trade, had played in the Knights' very first match in 1988. His loyalty to the club was unquestioned, and he was an effective sports administrator. Until late 2010, Tew had met Tinkler just once, a brief conversation at a Knights game in early 2009; it turned out he'd coached some of the guys Tinkler had played with as a young man.

The previous year, the club had gone to Tinkler cap-in-hand, and Patinack Farm had kicked in half a million dollars. But there was a misunderstanding: Tinkler had assumed it was a loan, to be repaid when the club was flush again, while the club believed it was a sponsorship deal, but that Patinack had never taken it up. At that stage, Tinkler had nothing he needed to promote to the public.

There had for years been speculation that Tinkler might buy the Knights, but his takeover of the Jets gave new impetus to the idea. In the week leading up to the announcement of the soccer deal, Tew met with Tinkler to discuss the possibility that the two clubs could be managed together. There were further discussions through October 2010 – once Tinkler even invited Tew and the Knights' CEO Steve Burraston up to his summer palace at Sapphire Beach. Tew asked the Knights' board of directors to think about the possibility of a privatisation, and to sound out close confidants.

On 8 November Tinkler offered a reported sum of up to $10 million to buy and operate the Newcastle Knights over the next 10 years, although it wasn't clear if that was per annum or a one-off sum, and to pay off all the club's outstanding liabilities. There was a key sticking point: Tew wanted Tinkler to put up a multimillion-dollar bank guarantee, so that if his business went broke or if he died, the club could call on some emergency funding to tide it over until a new sponsor could be found. Tinkler refused. Tew told the *Daily Telegraph* that he would not be prepared to put Tinkler's offer to members in its current form:

> With no guarantee Nathan will at least cover our existing revenues, I don't see how it puts us in a superior financial position to the one we are in now. I've conveyed to him that we don't think the offer goes far enough but I'm not sure if that is the end of it or not. The offer only arrived last night and I've had further correspondence today but that is as far as it has got.

The same day, Tinkler sent Tew an email; it was published some months later in the *Telegraph*:

> Hi Rob,
>
> As discussed previously I can not accept many of the headline issues which you outline ...
>
> I thought my offer was incredibly generous and would appeal to the community at large but you are too good for me.
>
> I will have Troy work with Steve on the forgiveness of the outstanding loan as my final gesture of supporting the club. There is nothing more I can do.

The public reaction to Tew's concerns was immediate: were the Knights mad, knocking back Tinkler? Coming out strongly in favour of Tinkler's proposal, the New South Wales Minister for the Hunter, Jodi McKay, attacked Tew and Burraston for apparently failing to take it to the board: 'If the directors did not know about this bid until after the decision to reject it was made, I do have grave concerns about what else they have not been told,' she said. The momentum was swinging Tinkler's way.

Six weeks later, just before Christmas, Tinkler invited Tew to another meeting, and a handshake deal was done. Tinkler's sweetened offer was a package worth up to $100 million to the Knights over 10 years, guaranteeing annual sponsorship revenues and including payment of the club's existing liabilities, up to $3.5 million. Another $2.5 million a year would be kicked into junior development. The Knights would continue to be run as a non profit entity – all revenues would go back into the club.

On 17 January the board came out in favour of the revised proposal; Tew called it simply 'too good to refuse'. A one-month deadline was set for the terms to be documented, so they could be put to members.

Around this time, Tinkler began acting as if he already owned the club. He made noises about wanting to hire Wayne Bennett, the legendary supercoach of the Brisbane Broncos and Queensland's all-conquering State of Origin sides. Bennett had engineered a massive turnaround at the St George Illawarra Dragons but was considered unlikely to stay there much longer.

In mid-February the Cronulla Sharks were left red-faced when their rising star Kade Snowden failed to show up to a press conference to announce that he was re-signing with the club for another two years for $300,000 a year. The club had been gazumped, revealing that Snowden, a former Knights player and junior at Belmont North, had received a counter-offer from a third party who'd called him an hour before the conference, and was now considering his options.

Tinkler was immediately fingered in the media. Snowden, an aggressive prop forward and a homegrown Newcastle talent, was said to be his favourite player. Apparently, Tinkler had offered the

24-year-old $2 million over four years – more than triple the Sharks' offer. Snowden duly reneged on his agreement.

As chairman of the Tinkler Sports Group, Ken Edwards officially denied that Tinkler had made any approach to Snowden, while Burraston said that an earlier Knights offer had been withdrawn more than a week before: 'We've had no negotiations with Kade Snowden and his management since that time. Any offer that came in recent times has not come from the Knights.' It was soon confirmed, however, that Tinkler had indeed called Snowden, breaking every rule in the book. It was unprofessional and unfair, dirty tactics – and Tinkler didn't even own the club yet! The NRL made noises about an investigation, but nothing came of it.

The Snowden incident put Tew on his guard. When Tinkler's deal documentation arrived in mid-February, the day before a press conference scheduled to announce that commercial terms had been reached, he was shocked to find key commitments missing. After getting legal advice and board approval, Tew spelled out these omissions in a press release.

Tinkler blew up, firing off an email: '[The Newcastle community] deserves better than it has been getting and I hope you can apply as much energy to moving it forward to deliver better outcomes as you have to holding it back in this process. Tew and Burro rejoice, you keep YOUR club.' For a second time, Tinkler walked away – and this time, he was not going quietly.

The key misunderstanding, it seems clear in retrospect, was over Tinkler's much-touted $100-million offer. Was it a commitment to put in $10 million a year for 10 years? Or just to top up any shortfall, if the Knights' own revenue was below $10 million a year? Was it a sponsorship? Or were hospitality and ticketing revenues also to be counted? By 2010, the Knights were attracting more than $7 million per year in sponsorship, including a hefty $1.2-million package from Rio Tinto's Coal & Allied – one of the most lucrative sponsorships in the league. If that revenue continued to pour in, Tinkler would only have to chip in $3 million a year. If sponsorship revenue jumped to $10 million, Tinkler might put in nothing. If hospitality and ticketing revenues were also included, the Knights were already pulling in more than $10 million a year, easily.

At a press conference that afternoon, Tew claimed that Tinkler's documented offer was a 'pale imitation' of the January deal, and accused him of moving the goalposts. 'Put simply, the much-talked about $100 million offer turned out to be a mirage,' he said 'The real value of the offer ultimately put on the table is a year-by-year commitment of between zero and $10 million paid over 10 years with a maximum guarantee of two years.' Burraston added that there was a lot of confusion and misinformation about the bid, but those who had seen Tinkler's actual document were 'very positive in rejecting it … we're disappointed too that it's not a $100 million offer, but it's certainly not.'

Tew and Burraston were backed up by the Knights' official supporters group, which said it was the board's responsibility to safeguard the club for members. Tew told Burraston to load up the offer documents on the club's website so that all could see what was missing from the offer.

What Tew couldn't say was that, amid a front-page debate about the future of the Knights, he was getting calls from people, some of whom he'd never met, tipping him off to what Tinkler was really like. In a tight community like the Hunter Valley, there are not six degrees of separation, just one. Friends of friends, known well enough to get Tew's phone number, would call to relate their own personal experiences with Tinkler. They would not go public; they did not want to make a fuss. They were not campaigning against the privatisation. They just wanted Tew, as the Knights' chairman, to know who he was dealing with. In a quiet Hunter Valley way, Tew was being put on notice. It stiffened his resolve.

For his part, Tinkler claimed that Tew and Burraston had misunderstood the original deal, and that the terms had never substantially changed. He went on the attack in a vintage spray, full of personal accusations, published the next day by the *Daily Telegraph* as a sensational interview. Tew and Burraston were playing 'real-life fantasy league' and were the only two people making decisions at the club, while 'everyone else is in the dark'. The transaction was apparently 'beyond them': 'they appear to be solely focused on disaster scenarios. I am surprised they dare go outside.' Implying they were only looking out for their own jobs, Tinkler said his mistake was 'not guaranteeing Tew's and Burraston's future involvement at the club'. It was an outrageous slur, given Tinkler had offered both men work at the club after the takeover but had been rebuffed.

For the first time, Tinkler also admitted that he had called Kade Snowden. He defended his actions and explained why the Knights needed an enforcer, saying: 'I didn't tell anyone. I don't need permission to use a phone and local talent should always be first priority.' Tinkler denied having made Snowden a counteroffer: 'I just asked, what's the hurry? I didn't realise he was being pushed to sign a new deal and had no clue the Knights had walked away from signing him. Kade should wait till the end of the year. His price is only going up.' It was as though Tinkler was the only one who could spot local talent, or who was committed to keeping it. 'We have established a strong culture of beating ourselves when it comes to developing talent,' he complained. 'Andrew Johns playing reserve grade straight out of school might have been pushed out of the club under this management.'

Tinkler also admitted he wanted Wayne Bennett to coach the Knights:

Who wouldn't? Any coach needs cattle and I would not be confident Wayne would take this squad on. He is a master coach and a master at identifying talent. You can't make a silk purse out of a sow's ear though and while coaching the likes of Kurt Gidley must be appealing to him, I am not sure he would be attracted to the current roster. Why take that on when you are a legend of the game whose legacy will be talked about for generations and you have your pick of nearly any club in the competition? Having said that, I think I could be very persuasive with him!

'We have wasted enough time,' Tinkler concluded. 'I think the whole thing is very unfortunate. Let's get on with running 10th! Looking forward to the season.'

It was over-the-top stuff, and unnecessarily nasty, but the punters lapped it up. That night, at the Mark Hotel in Lambton, not far from the Knights' headquarters, member Mark Ford started a petition to call an extraordinary general meeting to overthrow the board. In two hours, he got 57 of the requisite 100 signatures. Emotions were running high. The members clearly wanted their say. On Facebook, someone anonymously set up a 'Knights fans against Steve Burraston' page, which soon had postings such as 'Tew = clown' and 'get rid of these arrogant pricks in charge and start again' and 'here is the sword … now fall on it'.

Andrew Johns himself weighed in, attacking the Knights' board for blocking Tinkler's plan to foster more local talent: 'He's doing this for the community, for Newcastle and the Hunter. Why is it so hard to believe? … what the hell are they doing? We don't want Greg Bird playing for Gold Coast. We don't want Kade Snowden at Cronulla. We don't want Dane Tilse playing for Canberra.'

In a tough piece the next day, Phil Rothfield in the *Telegraph* disputed this, writing: 'Tinkler's wallet wouldn't have saved any of them.' Bird had left the Knights in 2001, when they were focused on keeping their Grand Final heroes, including Johns. Snowden had left because Knights coach Brian Smith didn't want him. Tilse had been sacked after a drunken incident with a female university student in Bathurst in 2005. Rothfield also pointed out that Johns had recently left the Knights coaching staff after a payments row.

Rothfield quoted NRL chief executive David Gallop, who had told journalists the previous day he had no fears over the financial future of Newcastle, despite Tinkler withdrawing his bid:

> It is perfectly reasonable to run a fine tooth comb over a proposal that would see ownership of a club passed over from the members to a private individual – that's a massive step. The Knights are loved by all Novocastrians and they are not on the brink of financial trouble. In fact, with the new grandstand and new television money coming up, the club's financial position will only improve in the years ahead. I have spoken to Burro in the last 24 hours. They are not opposed to private ownership but are within their rights to be looking for a deal that warrants such a massive change.

By the next Wednesday night, Tinkler had put his offer back to the Knights, with the missing terms written back in. Tinkler's side blamed poor drafting by their lawyers. Mark Fitzgibbon, the chief executive of health insurer NIB, which was a key sponsor of both the Jets and the Knights, stepped in as an honest broker, and both sides, bruised and battered, came back to the negotiating table.

The Knights weren't quite the financial basket case everyone was making out. There was constant reference to the club's $4 million in debts. That was overstated. There had been a gradual accumulation of losses over the years, held as a liability on the club's balance sheet, which stood at $2.3 million by 2010.

Firstly, however, that liability was from the club to itself – it was not as though there were creditors pounding at the door – and would only crystallise if for some catastrophic reason the show stopped completely and the club had to refund the next year's season ticket holders, who were making up the shortfall each year. Secondly, the club had returned to profitability over the past two years, and was gradually reducing that liability.

The losses had been exacerbated by a shocking season in 2007, when the club lost $1.3 million after an injured Johns was forced into early retirement, which had hurt the Knights' performance and ticket sales. The revenue slump was made tougher in 2008–09 by the demolition and construction of the western stand – traditionally the best seats in the house – which halved the club's potential ticketing revenue for two years. From the 2011 season onwards, the grandstand would be back in full operation, as Gallop had pointed out.

Arguably, the club was on the verge of a turnaround, and it had just signed a new back-of-shirt sponsor, Westrac. There was even speculation that Coal & Allied would drop its front-of-shirt sponsorship if the Knights were privatised. Lastly, the $4-million liability figure being bandied about included around $1.2 million in contingent liabilities, which might or might not ever be paid.

The largest of these was Patinack's 2008 half-million-dollar 'loan', which Tinkler was now demanding be repaid, having reneged on his offer to forgive it as a gesture of support for the club. Interestingly, that figure was reduced by some $40,000 because it turned out that Tinkler's Custom Mining had failed to pay the Knights for a corporate box which it had occupied in 2005. The Knights had tried to wind up Custom Mining at the time; the current board had realised only recently that Custom was a Tinkler entity!

Most importantly, to address a short-term financial problem the privatisation proposed a permanent transfer of a prized asset owned by members to a Tinkler company, which, if it kept its side of the bargain, could do whatever it wished with the Knights after 10 years. At least, under the status quo, if the members didn't like the way the club's assets were being managed, they could vote against the board, call a general meeting and even stand for election themselves. Those rights would be lost forever if the club was taken over by one of Tinkler's private companies, whose shares the members could not buy.

The Knights were not, of course, just a financial asset – a set of revenue streams set against a cost base – but an important piece of Newcastle heritage. Would Tinkler change the name? The colours? Sell off the team? Change it from being a non-profit to a for-profit organisation? To address these concerns, safeguards were built into the privatisation deal. The Knights' members, who were selling the club, would retain a single 'heritage share', which had special voting rights: 'heritage matters' – the name, the colours, the location, the home ground – could not be changed without them voting in favour.

A board elected by the members, and owning a single share in the Knights, would also be empowered to call on the bank guarantee in certain circumstances. The most important were if the Tinkler Sports Group, or the wholly owned subsidiary that would own the Knights, a company called Hunter Valley League Operations, had an 'insolvency event' – defined as including the appointment of a receiver or administrator, or the issuance of a wind-up order.

The problem with this structure was that the members, who would have no income or assets after the sale of the Knights, were effectively powerless to enforce their rights against Tinkler. And if the Tinkler entities became insolvent, the members would line up along with other creditors. This was clearly acknowledged as a risk in the explanatory memorandum put to members, along with the notice of meeting, which noted that the probable remedy would be to seek a court order for specific performance of the obligations on Tinkler (or an injunction to prevent a breach of those obligations). In all likelihood, the directors of the members' club would have to fund any such legal action personally – an extremely uncomfortable situation.

Recognising these problems, a few of the Newcastle business community, led by Andrew Poole, proposed a 'patron's trust' as an alternative to full privatisation. It was a purely philanthropic model: the club would remain the property of the members; $6 million to $10 million would be donated over four to five years, no strings attached. With all Tinkler's antics and outbursts, Tew and Burraston relied on the patron's trust as their fall-back plan, and put on a member's information night to explain it. Full-page ads were taken out in the *Newcastle Herald*.

Stephen Barrett, a former colleague of Poole's, published a blog, 'My Knights for Members', which pointed out the failings of the Tinkler bid, and posed a few questions about rumours of Tinkler's unpaid debts in the racing industry. By now there were all kinds of theories around, including that the Knights bid was a stepping stone for Tinkler to build a Newcastle casino. Tinkler threatened defamation action and got his new security firm, Internet Security Watchdog, which worked in online reputation management, to block the site.

It was part of an all-out PR offensive that was directed by Tinkler's spokesman Tim Allerton – who had advised Crowe and Holmes à Court on their bid for South Sydney – and backed up by an advertising blitz in the local print and television media. Ken Edwards was assiduous in cultivating local journalists, consulting with stakeholders such as the Knights supporter group the Crusade, and freely engaging with opponents to the privatisation. Anyone who wrote a letter to the local paper about the Tinkler bid got a phone call to talk through their issues.

Inevitably, as part of the sell-job, a lot of pressure fell on Tinkler – to explain himself, to answer the public's doubts, to show himself. In early March, Tinkler gave his first and last substantial TV interview, to NBN, the regional affiliate of Channel Nine. Sitting around a bland board table in a nondescript office in an open-necked shirt, he chatted freely with the legendary NBN sports reader Mike Rabbitt, a familiar face on Newcastle television since the early 1980s, about his upbringing, his trade and footy. Tinkler was utterly convincing: self-deprecating, witty, full of love for the Knights and the Hunter. Describing himself as an 'average Joe' whose family was still his 'number one asset', the multimillionaire suggested that the wealthier people are, 'the taller the tales are … I'm still the guy that a lot of guys in the Valley worked with and know and played football against.'

Tinkler acknowledged that he liked to keep a low profile and avoid the never-ending media 'circus', but that was partly because 'I haven't got a lot to say so there's no point'. He continued: 'It's important, I think, for members to sort of see me and understand that I am … doing this for the greater community … it's not a rich guy trying to force his way in, to own a football team.' It was all about the Knights:

> I'd love to see the Knights on the front page of the paper for winning games … Unfortunately, too much of the press over the last few years has been about what sort of disaster is off field, or the financial commitments of the club, or that sort of thing. That makes it very tough to win. You know, that takes a toll on everyone, coach, players, management, the whole lot … it doesn't have to be like that. I don't have the time or the inclination to run a football club or be involved in a hands-on sort of role. But what I can do is sort of lend a balance sheet to the whole thing, to make it run a lot smoother.

He acknowledged that there had been plenty of people before him who had helped the Knights out, but said the club was in a kind of spiral: it was 'always limping through under that sort of scenario … you can only go on for so long like that. If you want to compete and sort of collect premierships, I think you need to be fully funded.' Tinkler promised to respect the Knights' heritage:

> … the whole thing about, oh, you know, Tinkler will move the club and he can take it here and he'll change the colours and he'll do this … that's not what it's about. Those sort of things are locked away, you know, they're in a vault and they can never be changed, you know, unless the members want them to change. So, you know, it's not driven by me; they're not decisions for me to make. I'm just the schmuck putting the money in!

He was conciliatory towards Tew, saying he had a 'thankless task'; there was no doubt 'both of us had the best interests of the club at heart':

> Rob's worked tirelessly for Newcastle … I do this sort of stuff every day … Rob's probably been out of his comfort zone … I might have been a little bit impatient with that, but, ah, you know, we've both kept coming back to the table.

Again he went over the Snowden incident:

> Kade's a talent and I was shattered when he left. I was actually in New York and I got up and I seen in the paper that he was, ah, he'd re-signed. I rang Ken and I said, 'What happened here?' He said, 'Yeah, I dunno.' So I rang Kade and … I didn't realise he was about to front a press conference! But look, I

didn't make an offer. I just said, 'What's your hurry, you know? You know, the Knights are going through, we're going through a transition period, you know, what are you gaining by signing up now as opposed to June?' So you know, I didn't ring up and say, you know, 'I'll give you a million dollars tomorrow if you sign' or anything silly like that. I think there was a few Sydney journos that are Cronulla fans that ... had the knives out over that! But, ah, you know, it was just an entire beat-up. But, you know, it goes to the quality of the guy ... he knows what he wants, he wants to come home ... it took guts for him to do that.

He wasn't going to interfere in selection:

If you love rugby league, you've all got your opinions ... that's what barbecues are for. But, um, look, you know, Rick Stone had introduced me to Kade at the end of last year ... I think he's got green and gold all over him. It wasn't me saying 'I'm going to buy my favourite player'.

*

On 4 March 2012, the Knights board finally succumbed to the pressure and backed the Tinkler Sports Group's bid, resolving to put it to members at an extraordinary general meeting to be held at the Newcastle Entertainment Centre on Thursday 31 March. A 14-page explanatory memorandum was circulated, along with a meeting notice and proxy forms. It was on.

When the memorandum came out, Stephen Barrett checked and discovered that the Tinkler Sports Group had failed to register the business name Newcastle Knights Members Club – which was already listed in the memorandum as the entity that would own the heritage share. Barrett registered it himself, for $75, just to be a nuisance and to show how disorganised Tinkler's people were. By the time they realised, there wasn't time to reprint the memorandum, unless they rescheduled the meeting, so they had to negotiate with Barrett. The conversation went something like this:

'What did you do that for?'

'To show what fuckwits you are.'

'We'll run you down in the press.'

'Knock yourself out! I'm nobody! You think about what it'll cost you to hold another EGM and get back to me.'

Apparently, a small amount of money changed hands.

Although the board was now recommending the Tinkler privatisation, there were still some real grounds for concern, as Burraston acknowledged a week later in an interview with Foxtel in the lead-up to the meeting:

> In all deals there is risk. And there is risk in this deal. But [the directors] believe the commercial aspects of the deal, in other words the financial income that may come from it, outweigh those risks. And they're taking Nathan at his word that he's going to do the right thing by the club.

Burraston also cast doubt on whether the club actually needed a financial lifeline:

> We're essentially a break-even club. Over 23 years, we're less than two per cent on the wrong side of the ledger. That's not an ideal position – we'd all like to have more money – but we all operate under a salary cap, so there is only so much you can do with the money anyhow.

Tinkler was furious, and Burraston was told to pull his head in.

From then on, Tew would be the Knights' only spokesperson on the privatisation.

On the Monday before the historic vote, Tinkler himself took the stage at the Newcastle Panthers club to answer questions from members. Hundreds turned up, and Tinkler, who was wearing a suit but no tie, seemed nervous at first. He didn't need to be: they were already on his side.

The members booed him good-naturedly when he confessed to having once been a St George supporter. They laughed when he talked about his tough negotiations with Tew: 'Rob wanted to talk about me going broke all the time; that was kind of hard to take. And then we had to deal with the whole issue of my death and that was pretty tough to take also.' One wag asked how many premierships Tinkler was prepared to guarantee for the Knights over the next decade, and how many free beer tents he was going to build for members. He played a straight bat to both requests. It was a hero's reception, and finished with a standing ovation.

At the same venue the next night, when Poole got up to explain the patron's trust proposal to a much smaller crowd of around 60, there was downright hostility – even though Poole wasn't proposing to take their club off them, just to help fund it.

It all came to a head on the Thursday night, when a thousand members gathered at the entertainment centre to cast their votes. Tew got straight down to business, and the whole meeting lasted less than 45 minutes. The two resolutions that would give effect to the privatisation needed 75 per cent approval. Ahead of the meeting, the mood was clearly in favour of the bid, but both Tew and Ken Edwards, chairman of the Tinkler Sports Group, feared a close count, which would mean a divided club with a disaffected minority.

They need not have worried. Of about 3000 eligible Knights members, some 2300 voted, and the result was overwhelmingly in favour: the count showed that 97 per cent of members had backed Tinkler. When a clearly relieved Tew announced the result, the crowd broke out in the spine-tingling chant: 'Newcastle! Newcastle!' Edwards thanked the members, saying TSG was 'humbled' by the historic vote. Tinkler himself was nowhere to be seen.

Burraston quit almost immediately, saying the club had been through a 'torrid time of late and with new ownership it needs to be given a fresh start'. To downplay any suggestion that the takeover was ego-driven, the Tinkler Sports Group was renamed the Hunter Sports Group.

A fortnight after the vote, Wayne Bennett announced that he'd signed with the Knights for four years, starting in 2012. Tinkler had won a bidding war against Russell Crowe, who was desperate to secure Bennett to help break South Sydney's 40-year Grand Final drought. Tinkler's offer was reportedly extraordinary – a million-dollar sign-on fee, $1.5 million a year, plus bonuses of half a million dollars if the Knights made the finals and a million dollars if he won them a premiership. One factor that apparently swung Bennett towards Tinkler was that he'd be able to use Tinkler's private jet to visit his family in Brisbane, or to fly them down. In a prepared statement, Bennett said:

> I sat down with Nathan, and was struck by his passion and knowledge of rugby league ... Private ownership appeals to me. I like it. This guy wants to take Newcastle to another level. He wants our game to capture the imagination of the Hunter like never before, for the Knights to be the benchmark of rugby league in this country. That is the type of challenge that appeals to me. The Knights have already won two premierships, and have been a great club despite struggling with their finances. And that's where Nathan Tinkler comes in.

Tinkler's vision was coming to pass. There was just one problem: he was struggling to complete the takeover.

The settlement was meant to take five weeks, but it would end up taking four months as it became clear that Tinkler – supposedly worth more than $660 million – could not fund the $20-million bank guarantee. The problem, unsurprisingly, was that all the banks Hunter Sports Group spoke to wanted to see actual cash before they would give the guarantee. Tinkler didn't have the money. Instead, Tinkler's man Troy Palmer started flinging accusations about undisclosed liabilities on the Knights' books, which apparently had been hidden by the directors.

The Knights were put in a difficult position. 'We're trying to play hard-ball with Santa Claus,' a club source told one paper. 'We have no option but to give Nathan Tinkler as much time as he needs to get this deal done.'

With the 30 June deadline for settlement looming, Tinkler needed Tew to agree to an extension. A string of creditors to the club had been waiting patiently since the members' vote. Tinkler wanted to handle them himself but Tew insisted on a $3.5-million, non-refundable deposit to pay them down.

Amid the arm-wrestling, and with his distrust of Tinkler increasing, Tew managed to insert a crucial provision into the agreement: the members' group, owning its heritage share, could force a buyback of the Knights for a dollar (and keep the balance of the bank guarantee) if Tinkler breached his funding obligations.

Finally, on 5 August, the deal settled. The bank guarantee had been provided at last. It wasn't long before word got out that the money which Tinkler had deposited with Westpac was itself borrowed. It was true.

*

Two months after completing the Knights takeover, Tinkler's experience with the Newcastle Jets turned sour. Marquee signing Jason Culina had to have a second operation on his injured knee; it was beginning to look as though he would never kick a ball for the Jets, but the Hunter Sports Group would still have to pay out his $2.7-million contract.

On 4 October 2011, four days out from the Jets' season opener against the Melbourne Heart, Tinkler summarily sacked coach Branko Culina. It was a case of shoot first, ask questions later. Tinkler suspected Branko had known that his son's knee damage might be permanent, and felt he'd been dudded on his marquee player. He was reported to be 'livid', saying 'heads will roll'. In a statement, Tinkler said: 'Jason's injury could mean the Jets are without their marquee player for up to three seasons – not a good result for the club, supporters, sponsors and players.' Tinkler wanted Jason's contract annulled.

For Branko, a lifetime's service to Australian soccer – as player, coach and father of a Socceroo captain – had ended in disaster, with his contract torn up, amid bitter accusations of nepotism. The Jets had been undefeated in their pre-season campaign, and Branko had represented the club at the A-League's season launch at Sydney Football Stadium the previous day. Straight after the launch, he met Tinkler at his Martin Place offices and was told his contract had been terminated, effective immediately.

It was like a recurring nightmare for Branko, who had been sacked three years earlier as coach of Sydney FC after a losing streak. Branko would not work at the elite level again, and now coaches at Rockdale, in Sydney's south. He was replaced at the Jets by Gary van Egmond, who had coached the team to the championship victory in 2007–08.

It was a messy situation. Jason Culina's injury dated back to his last season with Clive Palmer's Gold Coast United team. His last game had been for the Socceroos against Korea in January 2011, just before he signed with the Jets, when he had come off at half-time. Jason's pre-existing injury meant the Jets could not insure him, and there was a question whether the club had done enough due diligence.

Branko had consistently said he was not responsible for signing Jason. He was backed up by the Newcastle-born soccer legend Ray Baartz, a former Manchester United player and member of the Jets' advisory board, who told the local *Herald* that Branko should not be made a scapegoat; he'd had to be persuaded to sign his own son:

> There have been some accusations around that Branko signed him and we bought a lemon and that sort of stuff about the father-son situation. It wasn't Branko's idea. The whole instigation of signing Jason came from the football advisory board and me in particular. I made the recommendation that he was the player we needed ... Jason is the best player in the league and will give us that X-factor. At the time, we knew he had a knee injury and we did our research on that ... at that stage there was no concern it would be a long-term injury. We signed a 30-year-old player who was the best player in the league, a current Socceroo and the medical advice we got was he'd had a minor cartilage operation with no concerns.

Miron Bleiberg, Jason's previous coach at the Gold Coast, however, had already gone on the record to say that Culina's knee condition was wear-and-tear more than a serious injury, and would likely persist for the rest of his career: 'It's nothing new in football, these young players with the body of 30-year-olds and the knees of 50-year-olds.'

In 2011–12, for the second time under Tinkler's ownership, the Jets ran seventh out of the 10-team comp. Behind the scenes, the goodwill between Tinkler and the club was fading fast. As former Jets adviser Neil Jameson told the ABC's *Four Corners* two years later, there was a mindset within the Tinkler camp that bordered on hubris, and it took a toll on everyone else. The early collegiate spirit evaporated and was replaced by an aversion to external advice:

> ... a sense of dogma started to appear whereby you had to adhere to the cult of Nathan, that Nathan was a genius, that what Nathan had achieved that no one else had achieved ... things had to be done his way or not at all. [It was] beyond arrogance. It's something I'd never seen before, and it was a little bit astounding because you couldn't break through it ... What we observed in the Jets and the Hunter Sports Group was that, initially, Nathan would bring people in and they'd be very warm and you'd be very welcome and there'd be a sense of love and belonging, but just as quickly that would cool and be replaced by a sense of disassociation and even bitterness. The romance cools and it cools dramatically, and that's not pleasant. You don't see it coming, and it's not pleasant to be part of it.

Tinkler turned on the FFA itself. Its chairman, Lowy, was in the middle of a public battle with the volatile owner of Gold Coast United, Clive Palmer, whose team had consistently had the worst attendance figures of any club in the A-League. Attacking the FFA publicly, the colourful mining magnate skited in an interview with SBS: 'We've paid $500,000 for our licence. Poor old Nathan Tinkler had to fork out $7 million for his ... I don't know why that happened.' In late February 2012, after Palmer sent his team out wearing 'Freedom of Speech' logos, he was stripped of his A-League licence by the FFA. Gold Coast United would be replaced by the Western Sydney Wanderers in 2012–13.

Palmer had his figures wrong, but nevertheless Tinkler was incensed: the clear implication was that he'd had his pants pulled down by Lowy. Two days after Palmer's outburst, Ken Edwards resigned as chairman of the Hunter Sports Group, ostensibly to spend more time with his family. More than anyone, it was Edwards who had delivered the Jets and the Knights into Tinkler's hands. But according to press reports at the time, when Tinkler started looking into what he had paid for the Jets, he was furious to find that Edwards had been paid a hefty bonus – somewhere between a quarter and half a million dollars – for luring Hunter Sports into a takeover of the club.

The FFA said Hunter Sports had known at the time that Edwards was working for the FFA and would be remunerated in the normal course of business. Edwards, being under a confidentiality agreement, made no public comment.

HSG demanded to know how the FFA had calculated the $4.5-million purchase price for the Jets. Former A-League chief Archie Fraser told media that the 'acquisition fee' of $3.5 million, which made up the bulk of the deal, had been 'invented' to extract as much money as possible out of Tinkler. No other club had had to pay it! The FFA denied that, but said the acquisition fee was unique to each club and reflected the market, its population, history and supporter base. Newcastle, with a decade's history and the third-biggest turnout in the comp, was simply worth more than the Gold Coast. For Hunter Sports, that didn't wash: it claimed it had been told the acquisition fee was standard. Plus, there was the Culina saga.

On 10 April 2012 Tinkler dramatically announced that the Hunter Sports Group would hand back the Jets' licence altogether, citing an irrevocable breakdown of confidence in the present FFA management, and walk away from its investment of some $12 million over 18 months. Tinkler slammed the FFA, which had just revealed combined losses of $27 million across the 10 clubs of the A-League, comparing it unfavourably to the NRL.

The Jets' players and supporters were stunned. It was unthinkable that Newcastle would be without a soccer team. The FFA hit back immediately, with CEO Ben Buckley telling the media there had been repeated attempts to negotiate with Tinkler and Palmer, and pointing out that 'a club cannot just hand back a licence … the Hunter Sports Group has a legally binding contract that goes through until June 2020 to participate in the A-League and field a team in the A-League. We expect them to fully commit and fully honour that contract.' The FFA was reported to be considering a $50-million damages claim against the Hunter Sports Group.

As he had offered to do with Palmer – with no luck – Frank Lowy flew up to meet with Tinkler at the end of April to resolve the impasse. They met at Brisbane airport in Tinkler's private hangar (which he'd bought from the Talbot estate). Lowy made unspecified concessions and the two men hammered out a peace deal.

'It was a good opportunity for Nathan and I to meet face-to-face,' said Lowy. 'When Nathan first took on the Newcastle Jets, I welcomed his commitment … today, my confidence is renewed over Nathan's personal drive to make the Jets and the Hyundai A-League successful.'

The old charmer had worked his magic. 'Frank Lowy flew to Brisbane to meet with me face-to-face and I took that as a sign of goodwill,' said Tinkler, and he reaffirmed his commitment to the club.

This episode was hardly reassuring, though, for the Jets or the Knights. Being owned by Hunter Sports was proving to be anything but smooth sailing. The day after the company tried to hand back the Jets' licence, former Knight Matthew Johns – Andrew's older brother – gave a prescient warning on radio station Triple M:

Knights fans have every right to feel concerned ... I wonder what Wayne Bennett is thinking this morning. Tinkler sacked Ken Edwards, who is a good friend of Wayne's. The Knights board were warned by some high-profile members about Tinkler, [but] at the end of the day the offer was, for the members, too good to refuse. The safety net for the Newcastle Knights and their members is the fact that they have a $20-million bank guarantee. There are concerns about what's going on and how he's conducting his business ... No one seems to know exactly what is Nathan Tinkler's financial position. I don't know if it's cash-flow problems, but you talk to people at the moment [and] everyone is going, 'We're not sure what is going on.' No one knows what Tinkler is going to do next. Nathan Tinkler has proved himself to be extremely volatile. He's a man with little patience, and even less loyalty.

*

Tinkler was not Newcastle's Santa Claus. Buying the Jets and the Knights, many thought, was all part of his plan to build a brand-new, $1.5-billion coal loader on the old BHP steelworks site, on the southern bank of the south arm of the Hunter River, five kilometres up from the centre of Newcastle Harbour.

Tinkler's business empire was expanding beyond coal and horses, into football, property and infrastructure. The developer Buildev, which he'd bought into in 2008, was a key plank of the strategy. Though Buildev was based in Newcastle, most of its projects were in regional New South Wales or Queensland. Tinkler wanted Buildev to connect with the local community: at one point, he even declared it would tackle the city's most intractable urban renewal project, the redevelopment of the Hunter Mall, which spans two blocks in the heart of the CBD. Buildev would buy the site from property giant GPT Group for $600 million. There was no follow-through, however, and Tinkler's grand plans unravelled within a few months.

Amid the expansion, the very core of Tinkler's vision was to have a 'vertically integrated' coal company, with operations from mine to port, and not beholden to the big miners like BHP and Rio. He wanted to run his own mines. He wanted his own train sets. And he wanted his own coal terminals.

For years, the port of Newcastle had been bursting at the seams. In the early years of the resources boom, the big miners, transport and logistics operators realised they needed to work together to plan future investment and make the best use of the rail and port infrastructure along the Hunter Valley Coal Chain – the world's biggest coal supply chain, which delivered 60 blends of coal from more than 30 mines operated by over a dozen companies, with just two weeks' demand visibility.

Newcastle's two big coal terminals were the bottleneck, and the ACCC was called in to allocate port capacity under a new competitive framework that ensured access for new players – effectively making sure that the big boys didn't run the port as a cartel. From 2004, all the talk was of doubling Newcastle's export capacity to 200 million tonnes per annum. The two existing loaders were expanded and construction got underway on a third, to be built by a breakaway group of miners, led by BHP, called the Newcastle Coal Infrastructure Group (NCIG). Those projects would take years to finish.

Meanwhile, in 2007 the queue of ships off Newcastle hit a record 79, stretching as far as the eye could see. It was blindingly obvious that there had been massive under-investment in coal infrastructure. Most people believed a fourth coal terminal would be needed by 2015, at the latest. Tinkler didn't blame the government for the under-investment; he blamed the coal industry, telling the *AFR* in late 2010 that it was not up to the government to build ports or rail lines, it was up to the industry to invest, and that 'warring between the mining companies is absolutely stifling it.'

At the same time, the New South Wales government's three ports strategy flagged that Newcastle would diversify to include a new container terminal once Port Botany reached capacity. So in December 2008 planning minister Kristina Keneally announced that the Buildev–Intertrade consortium had won a tender for a $120-million redevelopment of 62 hectares of *non*-waterfront land at the old BHP steelworks site at Mayfield. The tender criteria envisaged commercial and light-industrial uses – warehouses, distribution, that sort of thing. No one had imagined putting a coal loader on the site.

The tender rules didn't stop Nathan Tinkler, though, who in 2010 proposed constructing a new coal loader at Mayfield, a much higher-value use of the site. Apparently, he'd had a light-bulb moment at a meeting, after which the Buildev guys asked themselves, 'Why didn't we think of that?' Much later, they would come to believe that building a loader at Mayfield had been Tinkler's motive for buying into Buildev all along.

On 1 December 2010 the operator of two of Newcastle's terminals, Port Waratah Coal Services (PWCS) – majority-owned by Rio Tinto and Xstrata – announced its allocations, and Tinkler's Aston Resources got less than half of the export capacity it was seeking for its Maules Creek project. It seemed Aston's coal might be stranded, unable to get to market for the first three years, meaning millions of dollars in lost earnings.

Aston chief Todd Hannigan wrote an outraged letter to the ACCC, claiming that the incumbents were 'hoarding capacity'. After some industry consultation, the ACCC rejected Aston's request comprehensively, denying that there was any urgent reason to review the capacity arrangements.

Tinkler decided to go public with his bold plan. The day after the PWCS allocations came out, the *Newcastle Herald*'s Ian Kirkwood revealed that Tinkler hoped to use at least part of the Mayfield site to build a new coal loader. A decade's planning for the site, and a drawn-out competitive public tender process, would be thrown out the window. The *Herald*'s editorial acknowledged that a loader on the site was probably 'the last thing many would expect or want' but called for a proper appraisal of Tinkler's idea.

If it was to work, the land's owner, the Hunter Development Corporation, would need to approve Tinkler's loader, while the state-owned Newcastle Port Corporation, which controlled the 90-hectare waterfront section of the old steelworks site, would need to provide him with a berth. Both agencies were squarely opposed. The member for Newcastle and minister for the Hunter region, Jodi McKay, sided with a local residents' group and was 'categorically' against the plan. Yet Tinkler's spokesman said he had received a very positive hearing from the 'upper echelons' of the state's Labor government.

Increasingly alarmed, McKay watched as Tinkler's loader plan got traction inside the government – particularly, it seemed, within the office of treasurer Eric Roozendaal. The Buildev team was seen constantly around the State Parliament. Treasury officials were mindful of the previous tender process, and believed that the shortage of port capacity would be temporary (in fact, rail would soon be the constraint, not port), and they were also dubious that a container terminal at Newcastle would ever stack up economically.

Tinkler was trying to flex his muscles. In one meeting with McKay, ostensibly about his takeover of the Newcastle Knights, he took the opportunity to raise the coal loader proposal and offered to donate to McKay's re-election campaign. She refused: Tinkler was a developer, and any donation from him would be in breach of the electoral funding laws. Tinkler replied that he had hundreds of employees and could get around the rules. McKay again declined.

As the March 2011 state election approached, McKay feared that an approval for the loader would be rushed through before the government went into caretaker mode. When Treasury's advice refuting the case for a container terminal was leaked to the *Newcastle Herald*, she rang Roozendaal, screaming: 'I know what you're up to!'

Roozendaal answered: 'Haven't you spoken to Tinkler?'

By now at odds with the treasurer, McKay knew her time was up. She had fallen afoul of Labor's right-wing powerbrokers.

Days later, an expensive colour-printed leaflet opposing McKay's position on the issue was distributed to thousands of residents by a single mum and her kids, who had been paid for the work in cash. Circulated after writs for the state election had been issued, the unauthorised leaflet was a breach of electoral laws. McKay, suspecting that her Labor colleagues had betrayed her, took her concerns to the police, the electoral commission and, soon, to the Independent Commission Against Corruption (ICAC). In March 2011, as expected, Labor was annihilated at the state election. McKay lost her seat and was replaced by Tim Owen, the first Liberal to represent Newcastle in almost a century.

The election of a Coalition premier, Barry O'Farrell, and the appointment of his new ports minister, Duncan Gay, meant Tinkler had to refocus his lobbying effort. Within six weeks, Gay declared he had not ruled out a coal loader at Mayfield; all aspects of the former government's ports strategy were up for review. Tinkler's plans – still not yet lodged publicly – got bigger and bolder.

This time, rather than doing backroom deals with politicians, Tinkler set out to win hearts and minds in Newcastle. Hunter Ports became the main jersey sponsor of both the Knights and the Jets. In May 2012 Tinkler wrote to the residents' group, Correct

Planning and Consultation for Mayfield, thanking them for their invitation for him to brief locals on the development of Hunter Ports. Signed by Tinkler himself, the letter was particularly warm towards residents and hostile to the rest of the coal industry:

> I sympathise with residents of Mayfield and believe the efforts made to date have been nothing more than token from PWCS and NCIG. I am a proud Newcastle citizen and I am embarrassed at recent major developments such as NCIG and what they have done to our beautiful city ... The people of Newcastle were shortchanged significantly to the point where the state received a largely expanded coal loader and a completely new coal terminal; and the people of Newcastle got a black eye and new one lane bridge so that everybody gets a chance to slow down in the congested traffic on the way to the airport. This is a disgraceful result for Newcastle and one that Hunter Ports will not follow despite the bar being set very low for further development ...

Tinkler tried to position himself as the residents' champion. He proposed a new rail corridor, travelling along the river's edge, which would allow the closure of a kilometre-long stretch of rail that went straight through the suburbs of Mayfield and Tighes Hill. He claimed he would remove 90 per cent of coal trains from the local residential community.

Tinkler's coal loader would be state-of-the-art, setting new standards for management of dust and noise, with giant barriers erected to shield the stockpiles from the prevailing winds, fully enclosed conveyors and a covered, in-ground unloading facility. There were other environmental benefits too: there would be no damage to the riverbanks and wetlands upriver, and no need for a massive dredge to deepen the river.

On the flipside, Tinkler had doubled the capacity of his proposed loader, to 100 million tonnes per annum. The project cost jumped from $1.5 billion to $2.5 billion, and the proposal expanded onto Newcastle Port Corporation's portside land, which Tinkler hoped he might get for free from the state government under its new 'unsolicited proposals' process, which bypassed the ordinary planning regime and put proposals directly in front of the premier. It was exactly the same process that James Packer later used to win approval for his casino/hotel/apartment development on the Barangaroo waterfront at Sydney's Darling Harbour.

Tinkler next made an appearance at the Spring Ball at New South Wales' Parliament House, and was seen talking with Premier O'Farrell and Treasurer Mike Baird. He held meetings with Deputy Premier Andrew Stoner, whose portfolio covered regional infrastructure and investment. He letterboxed 14,000 homes in the suburbs around the steelworks site.

In October Tinkler sent project manager Steve van Barneveld to a briefing organised by Correct Planning and Consultation for Mayfield. Over 200 people turned up at the historic Mayfield East Public School. Hunter Ports was given a polite hearing – there was genuine appreciation for aspects of Tinkler's proposal – but the meeting passed a unanimous resolution: no new coal loader at Mayfield, and no more coal loaders for Newcastle. According to spokesman John Hayes, even in a city where every second person either works in the coal industry or knows someone who does, the mood of the residents was clear: they were over coal. There was scepticism even when Tinkler promised to set up a $20-million fund that would bring profits from the loader back into the community, from a per-tonne levy on coal shipped from the terminal.

There were more meetings and the letters poured in to the *Newcastle Herald*. What never eventuated from Hunter Ports was a publicly available, formal application which set out in detail what Tinkler was actually proposing to do. A few pretty visualisations handed out to residents were not going to clinch the debate.

Around this time, the commodore of the Newcastle Cruising Yacht Club, Phil Arnall, quietly took the new head of Infrastructure NSW, Nick Greiner, on a boat tour of Newcastle harbour. Arnall showed the former premier of New South Wales where Tinkler wanted to put his coal loader at Mayfield, and Greiner was unusually blunt: it would never happen. He would later describe it as 'bloody-minded nonsense'.

*

Sometime between 11 p.m. on Saturday 25 June 2011 and one a.m. the next morning, while Tinkler and his family were asleep at their Ocean Street mansion, someone snuck into the garage downstairs through a sliding door, found the keys to his black 2010 Ferrari California and drove off in it. Two days later, after a tip-off, police found the car torched in bushland at Raymond Terrace, near the Pacific Highway. It was a big story: national news bulletins showed pictures of the wreck, completely burned out, only its mag wheels hinting that the car was a Ferrari.

Tinkler, understandably, was spooked. His security consultant, Ken Gamble, who ran the firm Internet Security Watchdog, had given Tinkler advice on how to secure the Ocean Street property, but he had ignored it. From now on, security would be beefed up. Gamble placed three ex-SAS security guards on permanent assignment to protect Tinkler and his family: one to accompany him, one to accompany his wife and kids, and another to look after his properties.

A month later, a 17-year-old kid from Bar Beach was arrested and charged with five counts of aggravated breaking, entering and stealing, including of cars worth $200,000, committed in a spate of thefts in June. Detective Chief Inspector Wayne Humphrey said it was possible that Tinkler's Ferrari had caught fire accidentally:

> It was parked in long grass … I can't definitively pronounce [the accused boy] burnt the Ferrari, that's not to say he didn't intend to. It might have gone up in flames before he had the chance. Catalytic converters get quite hot; parked in long grass the heat generates flames – it's quite a plausible explanation.

The offender was sentenced to a year's juvenile detention; released after eight months, he was soon caught stealing more cars. By now 18, he pleaded guilty to the new offences and, according to the *Herald*, gloated how he'd taken Tinkler's Ferrari and driven it at speeds of up to 300 kilometres per hour.

It was an entirely random attack, but the rumour and innuendo would soon start: was it to do with the Supercar Club dispute? Was it some unpaid creditor? There were plenty of people, it seemed, who had a beef with the young billionaire.

7.

WHITEHAVEN
THE BOARDWALK TRANSACTION

In Brisbane, Aston's CEO Todd Hannigan just knew it: everything was going too smoothly. Aston's share price had risen steadily since its float in 2010, bolstered by the sale of a 15 per cent stake of the Maules Creek project to Japan's Itochu. By now Aston had refinanced its exorbitantly expensive loan from Farallon, paying it down to zero, which left the hedge fund and its investors with a stake of more than 20 per cent in the company. Planning was progressing, albeit slower than hoped, and Aston had enough money from its sell-downs to develop the mine. During 2011, contracts were awarded to build the coal handling and preparation plant, and the mine itself.

Rail and port access for Maules Creek remained problematic, and Tinkler and Hannigan were increasingly at odds over how to resolve it. Tinkler was pressing his Hunter Ports and Hunter Rail proposals at Hannigan. Tinkler had purposely stayed out of Aston because his next play was to be in infrastructure. From Aston's perspective, Tinkler's infrastructure plans were unfunded; what's more, Hunter Rail and Hunter Ports were related parties. In itself, that was not an insuperable conflict: the corporations law has mechanisms to manage related-party transactions, and the Aston board was free to consider the proposal without any direct influence from Tinkler, who was not a director. But Hannigan's duty was to act in the best interests of *all* shareholders, not just the largest. Tinkler lobbied Vaile to up the pressure on Hannigan – he seemed to have the mistaken idea that a company chairman could tell the CEO what to do. Hannigan had twice as many shares as Vaile – as did Tom Todd – and they were making decisions as both shareholders and executives.

Throughout 2011, even as Tinkler's coal loader proposal was progressing, Hannigan unveiled a series of deals: the first a take-or-pay deal with Newcastle's Port Waratah Coal Services, to load the first couple of million tonnes of Maules Creek coal in 2013, and ramp up from there; then a 5-million-tonne below-rail deal with the Australian Rail Track Corporation; then a savvy 7-million-tonne deal to use some of BHP's spare Newcastle Coal Infrastructure Group port capacity. By the end of the year Aston was in a position to export coal through both major terminal operators in Newcastle, and to blend any surplus product with other miners' coal for the first three years of operations at Maules Creek. The door was left open to Tinkler's coal loader, but Aston wouldn't rely on it alone. Hannigan knew this would aggravate Tinkler, and he was right.

*

Underlying the tension between Tinkler and Hannigan was a personal rivalry. In many ways, the educated, well spoken and presentable Hannigan was everything Tinkler was not. Hannigan could walk into a business meeting anywhere in the world and hold his own. The market loved him. In social settings, he mingled easily – he could even get away with the odd flirt. Tinkler was beginning to feel upstaged by his CEO, and it was eating away at him.

Hannigan was at pains not to steal the limelight, to allow Tinkler to take all the credit for Aston's success. He urged Tinkler to trust him, claiming to be delivering for him in spades, and asked to be allowed just to get on with it. But by now Tinkler was in falling-out-of-love mode, turning on his handpicked executive like he'd done many times before. As the year wore on, he became increasingly suspicious of his CEO, and uncommunicative.

Tinkler tried recruiting Tom Todd back out of Aston, to work with him on the unlisted side of the business and break the powerful alliance he had established with Hannigan. But Todd, now a rich man in his own right and holding a sizeable stake in Aston, could envisage a future without Tinkler. He preferred to concentrate on delivering Aston's potential, rather than join Tinkler on a new and riskier path, despite the promise of more equity and possibly greater riches in the end. Tinkler respected him less for taking the bird in the hand.

Tinkler relied increasingly on Philip Christensen, his man on the Aston board – and, some said, his attack dog. For a year and a half, as a director of Boardwalk Resources, Christensen had helped assemble for Tinkler a portfolio of coal exploration assets in New South Wales and Queensland. Their original aim had been to float Boardwalk for a motza; Christensen had equity in the company and stood to make a huge amount.

Boardwalk had interests in four coal projects – Dingo, Sienna, Monto and Ferndale – but they were a dubious collection, due variously to a lack of firm resource figures, regulatory obstacles and some dodgy partners. Dingo and Monto, both in Queensland, were transferred from Aston to Boardwalk for the princely sum of $10. Monto had anywhere between zero and 1.8 billion tonnes of coal – a range so wide as to be meaningless – and in fact the consultant Minarco considered the upper estimate misleading. Monto was an exercise in 'nearology': a punt that it would have a resource like Peabody's own Monto project, right next door, which was bogged down in an interminable dispute with QCoal.

Boardwalk also owned Sienna, bought from the listed miner Norton Gold in December 2010, but had no defined resource, and the permit area ran right over the town of Middlemount, making any planning approval unlikely, given the state's policy of ensuring buffer zones around urban areas.

Finally, Boardwalk had bought into the Ferndale coal project, near Denman in the upper Hunter Valley, and had also taken a 20 per cent stake in the project's listed owner, Coalworks. Inadvertently, Boardwalk had thus bought into New South Wales' most sensational corruption inquiry. Coalworks had bought the Ferndale project through Gardner Brook, an agent for the family of the infamous former Labor minister Eddie Obeid, who retained a 10 per cent interest. (Tinkler, the *AFR* later reported, had in fact met with Brook and Moses Obeid in late 2008, to discuss the upcoming tender of a mining lease at Mount Penny, the subject of the most sensational revelations at the 2013 ICAC inquiry. Through luck or good judgement, Tinkler had decided against investing in it.) Ferndale had the best defined resource of the four assets, however, and a Tinkler ally, Ian Craig, joined the board of both Coalworks and Boardwalk.

Altogether, Boardwalk had agreed to pay roughly $75 million for these four assets, half of which was still outstanding. It was borrowed money, of course. Farallon had lent Boardwalk US$50 million in January 2011, at an interest rate of 12 per cent for the first year, rising to 15 per cent in the second. As usual with Farallon, the debt came with warrants – effectively, free shares – in this case amounting to 20 per cent of Boardwalk. The interest clock was ticking and Boardwalk had little to show.

While Maules Creek was progressing well, it was still nowhere near production, and Aston was years from paying dividends. Aston's shares were going up – Tinkler's one-third stake was now worth close to $700 million – but his only way of accessing that money was to sell some shares or borrow against them. None of his other assets – such as Patinack or the Hunter Sports Group – was generating any cash flow. Unwilling to dilute his control over Aston, he chose to keep borrowing.

So, while Aston had been steadily paying down its debts to Farallon, Tinkler had been increasing his own. Even within the Tinkler Group, few realised exactly how much debt he was carrying, until an event in the middle of 2011 highlighted how fragile his finances had become.

As its shares had risen, Aston had grown fast enough to be included in the ASX200 index; when its inclusion was announced, Aston's shares dropped suddenly – from just under $10 to almost $8 – as fund managers rejigged their portfolios. The short, sharp fall triggered a margin call from Farallon to Tinkler, and he had to scramble to find some cash to pay down the loan secured against his Aston stake. If such a small movement in the share price could trigger such a near crisis, Tinkler was clearly already sailing close to the wind. Fortunately, in early August Peabody unveiled its $5-billion bid for Macarthur Coal, putting a rocket under the Aston share price, which headed for $12. Tinkler had a bit more headroom.

Tinkler's debts were mainly held in two private companies: Aston Resources Investments (ARI), which owned his 32 per cent stake in Aston, and Boardwalk Resources Investments (BRI), which owned his 83 per cent stake in the unlisted Boardwalk. Rebecca Tinkler was the sole shareholder in both companies, which were trustees for private family trusts.

With dozens of private companies and linked trusts, which together have generated hundreds of ASIC filings, it is difficult to trace the growth of Tinkler's overall debt to Farallon in this period, but it was clearly substantial. Aside from the US$50 million lent to Boardwalk, there were at least three separate Farallon facilities extended to Tinkler's companies that year. In April 2011, BRI borrowed a further US$130 million, and then another US$63 million in mid-October.

An investigation of Tinkler's finances by the *AFR*'s Angus Grigg and Jamie Freed would soon reveal that Tinkler had built up personal debts of up to $450 million by late 2011; he was starting to look 'stretched'. 'Very little of his empire is not mortgaged,' they wrote. 'Everything from his stake in Aston to lawn-mowers and X-ray machines at his Hunter Valley horse stud has been put up as security.'

It was a stunning story, and raised doubts about Tinkler's wealth – the Rich List figures were clearly wide of the mark. A fourth loan from Farallon – of US$23 million, used to fund his bank guarantee to the Newcastle Knights – took his debts to the Singapore hedge fund to US$266 million. In addition, there was a US$150-million loan facility from Credit Suisse, and about $40 million in loans from Westpac, secured against various property, including the Patinack stud farms. In July, according to the *AFR*, Tinkler had borrowed $4.07 million from Westpac, secured against farm equipment at his horse stud, including a 20-year-old Caterpillar grader, a Dixon Grizzly mower and a hay rake. He also pledged a 2008 Ford Focus and a Toyota Kluger.

Tinkler desperately needed to realise some value out of Boardwalk. One plan, codenamed 'Project Drifter', was to float the company. Most investment banks that looked at its portfolio of assets advised against it, but Morgan Stanley was willing to take on the assignment. It drafted a prospectus to raise $150 million for 100 million shares in the company, brazenly valuing Boardwalk at a staggering $750 million.

Tinkler was supposed to find four cornerstone investors to take most of the Boardwalk shares on offer, the draft said, leaving just 7 per cent of the shares for the public. Tinkler would remain chairman and keep complete control, retaining a 55 per cent stake worth over $415 million. Unlike the Aston prospectus, which was all about the asset, this time the investment pitch was all about Nathan Tinkler's track record of picking undervalued coal assets: he'd bought Middlemount for $30 million, and within four years it was worth $925 million; he'd bought Maules Creek for $480 million, and Aston Resources now had a market value of $2.3 billion. Investors were invited to imagine similar profits being turned from Dingo, Monto, Sienna and Ferndale. A timetable lodged with ASIC showed that the float was supposed to be done by the end of October 2011. But Tinkler could find no cornerstone investors to take up the fabulous opportunity, and the offer sank.

Tinkler did persuade the US coal miner Patriot to take a look at Boardwalk. He hoped to sell it to them in exchange for enough Patriot stock to make him the largest shareholder, then push it to bid for Aston. The plan foundered when, after due diligence, Patriot decided Boardwalk was worth nothing like what Tinkler was asking. A year later, Patriot itself went bust.

Meanwhile, in a genuine coup, in October Aston announced the sale of 10 per cent of Maules Creek to the Japanese utility J-Power, effectively valuing the project at $3.7 billion, eight times what Tinkler had paid for it two years earlier. It was a top-of-the-market deal.

With Aston suddenly cashed up, Tinkler tried selling Boardwalk into the company, but Hannigan was bitterly opposed. Aston had fought hard to pay off its Farallon debt, and was finally in a position to fund half a billion dollars' worth of development and construction at Maules Creek. Hannigan did not want to deviate from the strategy Aston had outlined in its prospectus by splurging the company's hard-earned cash on Boardwalk, which he saw as all but worthless. Nor did he want to issue new Aston shares in order to buy Boardwalk, diluting his existing shareholders and ceding majority control to Tinkler. Hannigan wanted to keep faith with the financial institutions that had backed Aston, knowing full well that he might need to raise fresh capital from them down the track. For Tinkler, this was close to betrayal. Aston was his baby, not Hannigan's.

It was not yet publicly apparent, but Aston's board was splitting beneath chairman Mark Vaile. Executive directors Hannigan and Todd were in one corner; Christensen and Vaile were in the other. Vaile owed Tinkler a lot. The $10 million-plus he'd made from his Aston stake was a fortune for an ex-politician, and he had no one else to thank but Tinkler. Personally, they were close; Vaile's daughter, who was studying at Newcastle University, even did part-time admin work for the Tinkler Group.

Former politicians are not guaranteed success in the boardroom; the short-lived premier of New South Wales Nick Greiner, a Harvard MBA who had a prolific career as a public company director after politics, is a rare example of one who made it. Vaile was a director of Virgin Australia, of the office provider Servcorp and of the Singapore-listed developer Stamford Land, but his only other chairmanship was at CBD Energy, run by another entrepreneurial figure, Gerry McGowan, founder of the former Impulse Airlines. CBD Energy was small-fry compared to Aston; if Vaile was going to establish himself in corporate Australia, he had to handle Aston well.

The emerging division on the Aston board put pressure on the former Sedgman chief executive Peter Hay, who had what counted for a long relationship with Tinkler, given Sedgman's extensive contracting work for Aston, and for Custom Mining before it. Like Vaile, Hay had made a small fortune from his association with Tinkler. Unlike Vaile, though, he did not feel indebted to Tinkler. He had helped Tinkler up from nothing. Now on his way to retirement after a long business career, Hay wanted a quiet life. He had played a key role in settling the nasty dispute between Tinkler and Ken Talbot's estate, and he had no illusions about the young billionaire.

*

Whitehaven Coal had been the underbidder when Aston had bought Maules Creek from Rio Tinto in 2009. It specialised in the Gunnedah Basin and was mining coal on either side of the Maules Creek deposit. The value of that project had been heavily underlined since the bidding war two years earlier, so if anybody was going to pay full value for Maules Creek – and even add some value – it was Whitehaven.

Never one to die wondering, Tinkler called Whitehaven's managing director, Tony Haggarty, to set up a meeting. Tinkler knew Haggarty, having approached him in 2006 as someone credible who might join the board of Custom Mining. Haggarty had politely declined – rightly or wrongly, he wasn't crazy about Middlemount – but he had watched Tinkler's subsequent progress with interest.

Going by the Rich List, Tinkler was by now worth twice Haggarty's wealth of $489 million. But Haggarty kept a low profile and had more than enough money to do what he liked. And after a lifetime in coal, he had something Tinkler could still only aspire to: standing. One of a 'gang of six' who had made their fortunes in coal many times over, and who'd stuck together over the years, Haggarty was good-natured, had the right industry connections – including with the all-important coal buyers in Japan – and had a proven track record in developing and operating coal mines. It was significant that when Tinkler wanted to discuss a deal with Whitehaven, he didn't call its chairman, John Conde – he went straight to Haggarty.

Haggarty was willing to discuss a possible merger. Tinkler had one condition: the merger must include Boardwalk; it was a package deal. Haggarty had no in-principle objection. For him, it was a question only of price.

It was no small transaction: a combination of Aston and Whitehaven would be a Top 100 company, with a market capitalisation of more than $5 billion, overtaking Robert Milner's New Hope Corporation to become the biggest pure-play coal company on the ASX. Tinkler got the Swiss bank UBS to work up some numbers, and the deal was codenamed 'Project Trifecta'. A meeting was scheduled at the Martin Place offices of the Tinkler Group. Haggarty and the independent Whitehaven director Andy Plummer – himself a significant Whitehaven shareholder, one of the old gang of six – were there on the Whitehaven side; Tinkler, Christensen and Vaile were there for Aston, and UBS's banker David de Pilla was the facilitator.

Aston's CEO Todd Hannigan had thus far been unaware that any merger discussions were taking place. Tinkler did not want Hannigan there – or Tom Todd, for that matter. It is unclear whether Haggarty and Plummer realised that the meeting was happening without the knowledge and support of Aston's whole board.

Mark Vaile was worried about going behind his CEO's back. He called Hannigan minutes beforehand to tell him he was going into a meeting with Whitehaven and Tinkler to discuss a potential merger. Hannigan was blunt: 'You can't go.' He challenged Vaile: how far had these discussions gone? Vaile insisted it was all exploratory and went in anyway.

Takeover talks have to start somewhere. There are no easy markers to indicate when an undocumented merger proposal becomes so well defined that it ought to be disclosed. Famously, when the chairmen of NAB and AMP met in a room at the Sofitel in Melbourne in 1999 to discuss a takeover, their chief executives were present. The controversy that ensued was not about the occurrence of the meeting but about the fact that AMP's shareholders were not subsequently given the chance to consider NAB's offer.

Naturally enough, Hannigan felt blindsided and immediately informed Tom Todd and Peter Hay. Both had also been kept in the dark. It was pretty clear that the writing was on the wall for all three of them.

The UBS term sheet proposed that Aston would buy Whitehaven via a friendly scheme of arrangement, with the declared support of the Whitehaven board and its key shareholders. The consideration would be a combination of cash and Aston shares – the cash component would come from a whopping $1.8 billion in borrowings. In a separate but simultaneous deal, Boardwalk would be bought with Aston shares for about $540 million. Tinkler would have three board seats, Whitehaven two; nothing was said about the chair or CEO. The proposal was presented to the Whitehaven board on 27 October 2011.

Aston's annual general meeting was on at 10 a.m. the next day, a clear Friday in Brisbane. The conference room at the Marriott was full of brokers and bankers, with a sprinkling of 'mum and dad' shareholders. A bitterly divided board took the stage, and Vaile and Hannigan went through the motions. One slide showed the strong 67 per cent rise in the Aston share price since the float, despite falls in the broader share market. Aston was the third-best performing stock in the Top 100. There was a minor protest vote over the remuneration report – in particular, over the rights being awarded to Hannigan and Todd – but all resolutions were passed on a show of hands, including the re-election of Vaile to the board. It all went perfectly smoothly, and nobody in the audience suspected what was going on behind the scenes.

Immediately afterwards, Tinkler called the executives to his Queen Street office to hear from UBS. The full board of Aston was present. Hannigan, seeing the UBS term sheet for the first time, started yelling: there was no way Boardwalk was worth $540 million; there was no way that Aston was going to be loaded with $1.8 billion in debt. Vaile, sitting between Hannigan and Tinkler, almost had to keep them physically separated. Tinkler's voice was lifting but he didn't explode; his mind was made up, and it would be embarrassing to have a barney in front of UBS.

Vaile, Hannigan, Christensen and Hay walked out to hold their board meeting at Aston's office down the road. When it came to Project Trifecta, Christensen was asked to leave: for one thing, he was Tinkler's representative, and for another he was conflicted, having equity in Boardwalk. Hannigan argued against the deal, which could do him out of a job. He convinced Vaile that Aston needed an adviser to act in its defence – in the interests of Aston's shareholders, not those of Whitehaven or Boardwalk. Vaile called his friend Trevor Rowe, chair of the local board of Rothschild.

That night all hell broke loose in the Tinkler camp. Why had the Aston board excluded Christensen? It was a risky strategy for any chief executive to challenge the company's largest shareholder. Tinkler and Christensen lobbied Vaile. For Hannigan, it could only end one way.

Tinkler was in the middle of a campaign against the board of Coalworks, whose shares had slumped since Boardwalk invested. Annoyed, Tinkler sided with Macquarie and others to vote down two of Coalworks' directors at its annual meeting in early November, and also voted down its remuneration report. Tinkler was blooded, and immediately afterwards texted Hannigan: 'I hope you saw that. They're gone. You're next!'

When Rothschild's report landed, it was devastating to the UBS proposal, without even directly attacking it. Rothschild showed that the proposed Boardwalk acquisition price was way over the odds when compared to the market value of any comparable coal explorer, or indeed to any comparable takeover transaction. No other float had occurred with such early-stage assets.

The assets had been bought for roughly $75 million, and the coal market had not lifted since they were bought; with no development occurring, how could such a large uplift on the purchase price be justified? Indeed, the outlook for coal prices was soft.

Each of the Boardwalk projects also had strategic issues. On a sum-of-the-parts analysis, Rothschild valued the assets at between $80 million and $149 million – but from that amount one had to deduct further acquisition costs of $36 million, and debts of $53 million. The net value of Boardwalk, they found, was between $2 million and $71 million – one-tenth, at the very most, of what Tinkler had originally wanted for it.

Rothschild's report also raised questions about the ability of the combined entity to service its debts of $1.8 billion. Aston was entering an expensive construction phase, while Whitehaven was still building its underground mine at Narrabri. On any normal commercial terms, Rothschild estimated that Whitehaven would struggle to service more than $900 million in debt.

Hannigan distributed the report by email and later that day got the call from Vaile: it was all over. On the morning of 17 November, a Thursday, it was announced that Hannigan and Tom Todd had resigned, with immediate effect. A courier was sent around to grab their phones and computers, and they were locked out of Aston's offices.

Tinkler called Hannigan and launched into a tirade: 'you're not going to get a fucking cent!' Hannigan just laughed, saying, 'Nathan, there's a contract, and Aston's going to honour that contract.' Both he and Todd had watertight agreements that included an explicit 'anti-Tinkler clause', giving them the right to resign in good faith if they disagreed with the actions of the major shareholder. The pair walked away with their entitlements fully intact, including six months' pay and share options worth more than $40 million each. Peter Hay resigned immediately, on principle. A fully costed deal with Pacific National, which would have solved Aston's biggest headache by providing above-rail access for the Maules Creek project, fell by the wayside, even though it had been previously approved by the board.

Tinkler installed himself as executive chairman, making Vaile his deputy in what looked like a demotion. An interim CEO and CFO were appointed from within the Tinkler Group. Vaile declared his support for the changes, and some nonsense spin was put to the ASX about the new skills Aston needed as it moved into its next phase. Aston shares plunged 11 per cent on the first day, from $10.75 to $9.55, as institutions including Colonial First State bailed out of the company altogether.

Analysts immediately suspected that Tinkler was trying to sell Boardwalk into Aston. Tinkler tried desperately to play down that angle, telling Freed at the *AFR* that related-party transactions were tough to execute and the possibility was not being entertained at the moment – which was complete rubbish. Aston's shares kept falling on day two as, suddenly, it was Tinkler fronting the presentations and fielding the questions. It was his first time at the helm of a public company, and institutional investors were hardly reassured. Tinkler said it was 'always planned that I joined the board of Aston' but that was news to the instos.

Curiously, when Tinkler was appointed to the Aston board, there was no disclosure that his shares were pledged to Farallon, although in the wake of the GFC there had been a concerted effort to tighten the listing rules to protect companies and shareholders against attacks by hedge funds on company principals who had large margin loans against their stake in the company. Under the ASX guidelines, newly appointed directors had to disclose any contract they had entered into which created a right to call on their shares in the company. Clearly, that applied to Tinkler's debts to Farallon and Credit Suisse, which were secured against his stakes in Aston and Boardwalk; had they been disclosed, much light would have been shed on his motivations for pursuing the merger.

Within days of the announcement, Aston's share price had fallen 20 per cent. As the documents were being drawn up, somebody leaked the negotiations to the *AFR*, and on Monday 5 December both Whitehaven and Aston confirmed that an indicative, non-binding merger of equals was on the table. That stopped the rot in Aston's share price. Rothschild was paid out its defence fee and its report was quietly shelved.

*

The full Whitehaven–Aston merger announcement was made a fortnight later, and had the unanimous support of both boards. Tinkler had wanted to remain as chairman, but after consultation with major shareholders and a bit of an arm-wrestle, it was decided that he would step down from the board. Vaile would chair the combined entity, while Haggarty would continue as MD. It aggrieved Tinkler no end that, once again, he was to be off the board of his own company. On the day of the announcement he was a conspicuous no-show on the media dial-in and investor briefings, leaving the explanations to Vaile and Haggarty.

The merger differed from what UBS had proposed in some key respects. For a start, it was no longer a takeover of Whitehaven by Aston but a merger of equals. There was no cash component – it was all scrip – and therefore no need for large borrowings. And this time around, the transactions were inter-conditional: it was a three-way tie-up or nothing.

The mechanics of the deal were fairly straightforward. First, Whitehaven would sweeten the offer by paying a one-off special dividend of 50 cents to its shareholders – roughly $250 million in cash. Second, Whitehaven would issue 1.89 new shares for every Aston share, valuing them at $10.05 each, based on Whitehaven's most recent share price. That was a small 3 per cent premium to the Aston price at the time, but was still below the $10.75 level the company had been trading at before the two Todds were sacked a month earlier. Third, in a significant concession,

Tinkler's Boardwalk Resources would kick in an extra $150 million to pay down the company's debts and to fund exploration. Boardwalk's owners would be issued 86 million Whitehaven shares as consideration for selling their company into the three-way merger, and another 34 million 'milestone shares' if any two of its four projects ever got regulatory approvals (no time limit was stipulated) or if a change-of-control event occurred, such as a takeover.

On the day the details were announced, most eyes went straight to the asking price for Boardwalk, which could effectively be as high as $537 million. The figure drew immediate fire from broking analysts: questions would be raised over the implied valuation, they said, there were no grounds for assuming production from any of Boardwalk's assets, and the acquisition price represented a 'significant, if not controversial premium to objective value'. Haggarty defended the proposed acquisition and explained why an independent valuation of Boardwalk hadn't been done. Haggarty did not know about the Rothschild analysis.

Market watchers would have to wait until March 2012, when an independent expert's report by PricewaterhouseCoopers came out with the scheme's explanatory memorandum (EM) for shareholders. By then, Whitehaven's shares had come back a bit: after the special dividend (and franking credits), they would be worth $4.58 each. Boardwalk's owners would be getting the same number of Whitehaven shares as before, but they were worth a bit less than when the deal was originally announced in December 2011: according to the EM, at least $393 million, and up to $491 million if the milestone shares were achieved.

But PricewaterhouseCoopers' report contained a bombshell: the independent expert had found Boardwalk was only worth between $200 million and $330 million. The merger had overvalued Boardwalk by at least $63 million, and perhaps by as much as

$291 million. If they weren't being fleeced, Whitehaven and Aston shareholders were certainly paying absolute top dollar.

PwC's valuation did take into account the $150-million cash contribution from Boardwalk. It had relied on valuations of Boardwalk's projects by the technical specialist Minarco – the same firm that had valued Maules Creek for the Aston float. Looking squarely at the projects, Minarco had come up with a valuation of Boardwalk between $50 million and $144 million, tops – roughly in line with Rothschild's valuation. The message coming back from the consultants was loud and clear.

Like many an underfunded explorer, Boardwalk was barely solvent. The independent expert's report noted that Boardwalk's liabilities had exceeded its assets by $5.6 million, as of 30 June 2011; if it couldn't raise more money, the auditors said, there was significant uncertainty over whether it could carry on. By December 2011, according to unaudited accounts reviewed by PricewaterhouseCoopers, that deficiency had grown to $14.5 million.

In most friendly mergers, 'independent' experts prove to be anything but – they consult to the proponents, and he who pays the piper calls the tune. Here, unusually, PwC had thrown a spotlight on a fundamental problem with the scheme.

The Boardwalk transaction was particularly disadvantageous for Aston shareholders. PwC calculated that they would get between $11.50 and $16.29 a share if they approved the merger with Whitehaven, but that this would drop to between $10.70 and $15.25 a share if the Boardwalk transaction was approved. This equated to a loss of between $164 million and $213 million on the deal, or a transfer of between 80 cents and $1.04 per share from Aston's shareholders to Tinkler and the other owners of Boardwalk. But Aston's shareholders were simply not given the option to approve the Whitehaven merger and reject the Boardwalk transaction.

In his letter to shareholders, Whitehaven chairman John Conde wrote: '... your board recognises that the value of the consideration payable for Boardwalk exceeds the valuation of Boardwalk determined by the independent expert'. The justification presented in the EM was a bald statement: 'The Whitehaven board has taken a commercial view of the tenements and the likelihood of their successful development under the Merged Group's ownership [and believes] that the consideration payable for Boardwalk represents fair value.' That was it. The board was simply saying: Trust us.

Proxy adviser CGI Glass Lewis told its clients – the institutional shareholders of Whitehaven – to vote no. When the scheme documents were released, one analyst – speaking off the record – told me: 'Somewhere in this document they have to convince Aston shareholders the Boardwalk transaction makes sense. That looks like it's completely missing. At some stage someone will have to come up with a valuation that justifies what they've come up with.'

Nobody did, however. Instead, the EM pointed to the 'synergies' that would be unlocked through the merger of Whitehaven and Aston. Consolidating the Gunnedah Basin operations would add value of between $575 million and $775 million, the two companies said, by sharing rail and port infrastructure, by allowing coal blending from the different mines in order to get better prices, and by enabling cost cuts on operations and overheads; all this should outweigh any concerns about overpaying for Boardwalk. PwC was less sanguine about the synergies – they came up with a range of between $510 and $680 million – but accepted the argument. Considered as a whole, the transaction was in the interests of shareholders, it found.

Undoubtedly, there were benefits to the combination of Whitehaven and Aston – defenders of the merger said it would prove the making of both companies – but every single dollar of added value could be wrung out of a two-way merger, without including

Boardwalk in the deal. On the independent expert's figures, Boardwalk only subtracted value from the merger. As Gavin Wendt of *Minelife* observed, the whole deal looked as though it was structured to solve Nathan Tinkler's financial problems. This was correct. Whitehaven wanted Maules Creek. To get it, Tinkler insisted it had to buy Boardwalk. It didn't want to, but it did.

It wasn't just analysts who were questioning the Boardwalk transaction. Long-time backers John Singleton and Mark Carnegie, holders of a significant number of Aston shares in their own right, were outraged at being diluted and opposed the deal behind the scenes. Their problem was that Farallon – which had brought them into Aston in the first place – was right behind the deal, which would crystallise the value of its Boardwalk warrants and give it better security for its outstanding loans.

The deal needed the approval of both Aston and Whitehaven shareholders, and this looked like it might go down to the wire. Despite declaring early on that they would vote for the merger, when the EM and meeting notices came out, the owners of Boardwalk – mainly, Tinkler and Farallon – decided to abstain from voting their combined 37 per cent stake, recognising that they stood to benefit from the Boardwalk transaction. 'It must be asked what caused such a dramatic change of heart,' wrote the *Australian*'s veteran columnist Bryan Frith, an expert on takeovers. 'It is suggested that the strong preference of the corporate regulator ASIC was Tinkler and Farallon should not vote.'

The merger required the approval of 75 per cent of the remaining Aston shareholders, at a meeting to be held in Brisbane on 16 April 2012, meaning that a protest vote above 15 per cent could scupper the transaction. The Whitehaven vote, by contrast, was almost a *fait accompli*: the share issue to Boardwalk required the approval of 50 per cent of Whitehaven shareholders, and the directors and their allies spoke for 43 per cent of the company.

The Federal Court judge Peter Jacobson, who was tasked with convening the shareholder meetings, pushed Aston's lawyers over whether Tinkler and his fellow Boardwalk shareholders were getting a special benefit from the Boardwalk transaction. As his later judgement explained:

> It might, on one view, be thought that the shareholders of Aston who also hold shares in Boardwalk could receive a collateral benefit, that is to say a benefit not available to other Aston shareholders, in the form of the consideration received from Whitehaven for their Boardwalk shares. This question arises because the value of the consideration, which Whitehaven has agreed to pay to Boardwalk shareholders, exceeds the valuation range for Boardwalk, which was assessed by PwC.

Justice Jacobson did note, however, that ASIC had not appeared to object, despite fair warning. He observed that Tinkler, Farallon and the other Boardwalk shareholders would not be voting their shares, and resolved to be practical. 'What seems to be important is that, according to the material to which I was taken in the scheme booklet, the Boardwalk transaction was struck as a result of arms-length negotiations,' he found.

Aston and Whitehaven had a green light – it was all systems go. If ASIC was watching, you wouldn't have known it.

*

On the voting day itself, Tinkler didn't bother to turn up – never mind that he was Aston's chairman. He had pre-existing business arrangements in Hong Kong, his deputy Mark Vaile explained. On a Monday morning, Aston's shareholders and hangers-on gathered in a room on Level 8 of the Brisbane Hilton to decide on the

$5-billion merger. For anyone expecting a bit of argy-bargy from the floor, or a protest vote, it was a fizzer. Vaile went into the meeting armed with a slide showing that the proxies already lodged were overwhelmingly in favour.

In the event, just six shareholders voted against the deal. But underlying this ringing endorsement was a strong message: this was a vote for new management. The investors wanted Haggarty in charge of Aston.

In a still auditorium at the Museum of Sydney, the board of Whitehaven was having a slightly harder time of it. Like Vaile, the company's urbane chairman, John Conde, held slides showing that proxies favoured both resolutions – on the issue of the shares to acquire Boardwalk, and for a doubling of directors' fees attendant on the merger – but this time the 'against' column showed significant protest votes of 10 per cent and 13 per cent. Worse, some pesky questions were asked by the half-dozen ordinary shareholders in the room, who took full advantage of their one opportunity to grill the board.

As often happens, their questions went straight to the heart of the matter. 'Don't take too much comfort in a high number of people voting "for",' warned one shareholder, Joy, who told the directors she had been one of the 3 per cent who, at the 2007 extraordinary general meeting of Rio Tinto, had voted against its disastrous Alcan acquisition – a deal that had almost wiped out the company, and had resulted in the forced sale of Maules Creek, the very asset Whitehaven was now buying. 'A 10 per cent vote against, that's three times as high,' she noted.

One man, Graham, was comfortable enough with the Aston merger, but concerned about Boardwalk: 'It seems to me we're a bit crazy paying so much for it, particularly in light of the independent expert's review.' Conde conceded that this question had 'exercised the minds of your directors' and threw to Haggarty, who suggested the independent expert had taken a 'somewhat narrow view of value'. The value of resource projects changed dramatically as you took them from one stage to another: 'One day it's zero and the next day it's a large number.' It was the nature of the business. Graham remained dubious.

After some debate about increasing the board's fee pool from $1 million to $2.5 million – a resolution rejected on a tiny show of hands, embarrassingly enough for Conde – Graham arked up again. 'Whitehaven's been a great company,' he said. 'We're just diluting our assets so much, and with a bigger board, that's going to cost us more. Why should we destroy such a good company?'

Conde took polite exception to this. It wasn't dilutive, he said, it was transformative, a step change.

'A step backwards!' retorted Graham, who pointed out that the market didn't seem to agree with the board, having sold Whitehaven shares down from $7 to $5 since the merger was announced.

But it was all too little, too late; the deal was done. The court approved the merger two days later and Aston was delisted from the stock exchange. Whatever concerns there may have been about the Boardwalk transaction were washed away. The whole deal had been heavily scrutinised over a five-month period and approved by an overwhelming majority of shareholders.

Four months later, when Whitehaven released its end of financial year accounts, one line item received little attention: a $120 million writedown of the 'fair value of Boardwalk Resources goodwill at acquisition'. There were some technicalities around the writedown, which reflected Whitehaven's view that the milestone shares would never be issued, but deep inside the accounts – a footnote to a note – was a plain admission: 'the consideration paid by the company was greater than the fair value of identifiable net assets acquired.' The acquisition was made on 1 May, with the Whitehaven share price at $5.18, and the company's 2011–12 accounts put the effective acquisition price at $495 million … *after* the writedown.

*

Nathan Tinkler, it seemed, had done it again. After failing to float Boardwalk and then failing to sell it, he had finally, through a tumultuous merger with Whitehaven and Aston, got half a billion dollars for a bunch of early-stage exploration assets which had so far cost Boardwalk just $39 million. What's more, another 34 million Whitehaven shares were there for the taking if the development milestones were ever achieved, or if a takeover occurred.

Setting aside the milestone shares, as a result of the Boardwalk transaction Tinkler himself received Whitehaven stock that was then worth $308 million. Farallon got shares worth $89 million, and Philip Christensen $15 million. The rest of Boardwalk's shareholders, who owned 7 per cent between them, split stock worth $32 million.

Those Boardwalk shareholders made an interesting list of the people who mattered most to Tinkler: there were his parents, Les and Zelda, and his sister, Donna; there were Rebecca's parents, Peter and Helen Blackney; there were his executives Troy Palmer, Matt Keen, Peter Kane, Ross Brims, Steven van Barneveld; there was his personal assistant, Sally Reynolds; there was John Thompson and Ben Lawrence from Patinack; there was Ian Craig from Coalworks. All shared in Boardwalk's good fortune: if the stock was evenly spread around, they would have received a couple of million dollars each.

Tinkler, for the first time, was the major shareholder in a liquid, ASX-listed company that actually mined coal and generated income. His 19.4 per cent stake in Whitehaven Coal was worth $1.1 billion.

Not that he was necessarily planning to hang around on the Whitehaven register. In fact, a number of commentators predicted he would do the opposite. There was abundant speculation that the merger was an attempt by both parties to bulk up and attract a bid from an overseas player.

In December 2011, belying the tussle behind the scenes about a board seat, Tinkler had hinted to Bloomberg he had 'put good people on the [Whitehaven–Aston] board to represent my interests'. In coal, these were increasingly offshore: 'I'm seeing a lot more opportunities overseas these days than in Australia and I want to be free to pursue those.'

The previous month – in fact, on the day before he took over Aston and sacked the two Todds – Tinkler had paid US$15 million for a mansion in Hawaii's exclusive Makena district. On a hill looking out over the Pacific Ocean, it was a huge home on more than a hectare of land in a gated subdivision one street back from the beach. While the best Makena properties can go for more than US$25 million, one specialist local high-end real-estate agent, Peter Gelsey, was amazed by the price Tinkler paid, telling me that properties 'across road from ocean' generally sold for less than US$8 million, especially since the Hawaiian property market had been devastated by the GFC. 'It appears Mr Tinkler paid more than double the typical rate for comparable properties in this luxury neighbourhood,' Gelsey said. The purchase raised questions about Tinkler's money flow offshore.

The house was bought in the name of his company Queen St Property Holdings. After a few hiccups when Tinkler apparently failed to settle, it was finally funded when he borrowed another US$16 million from Farallon in January 2012, increasing to US$39 million the loan he had used to fund the Newcastle Knights' bank guarantee.

Maui appeared to be a classic Tinkler acquisition: overbought, on a whim, with money he didn't have. It was precisely the kind of compulsive spending that stumped his advisers, especially Philip Christensen and Paul Flynn, the former Ernst & Young partner who had been Tinkler's auditor for years and had recently become chief executive of the Tinkler Group.

At a meeting in Tinkler's Brisbane office, it came to a head. Christensen and Flynn insisted that Tinkler rein in his spending, telling him that the outlook for coal markets was deteriorating and his asset values were falling, while his liabilities were rising. Already the company was out of cash, and the Farallon credit card was maxed out. Constantly batting away creditors, Christensen and Flynn felt their own reputations were at stake. Tinkler had to batten down the hatches.

They gave him a list of assets he should sell, ranging from easy (properties and jets) to hard (Aston Metals). Unperturbed, Tinkler wanted to keep buying: Peabody's Wilkie Creek mine in the Bowen Basin was on the market, as was Brazilian giant Vale's Integra mine in the Hunter Valley. Both were substantial operating mines, assets worth hundreds of millions. Tinkler wasn't listening.

His priority was buying a bigger, five-storey office building in Brisbane, at 443 Queen Street. Flynn had advised him squarely against it. Tinkler bypassed him and committed to pay $40 million for the building, in the name of one of his dormant companies, Francis Street Holdings, giving in-house lawyer Aimee Hyde and adviser Matthew Keen explicit instructions not to tell Flynn. Flynn's position had become untenable.

On 29 March 2012, even as Christensen and Flynn were about to join the board of the merged Whitehaven, ostensibly as his representatives, Tinkler sacked both men; CFO Mike Wotherspoon would join them soon afterwards as the entire Sydney office was shut down. It was dressed up as 'gardening leave' – Tinkler's spokesman told one reporter that Christensen had resigned and was on a 'well-earned break'. Both Christensen and Flynn were in fact legally employed by Tinkler until the end of September 2012, but it would turn out to be a permanent split.

It was now clear that Tinkler would not hear a negative word, even from his most trusted advisers. He called it 'Hannigan syndrome' – his own executives telling him what he could and couldn't do – and he hated it. At the end of that month, he pledged just about everything he owned to consolidate his debts with Farallon, taking out the biggest loan of his life, enough to make your head spin, and at crippling interest rates. The extra debt he'd strapped on to get his three-way merger away would prove the last straw.

Within days of the merger, Whitehaven's share price started falling and, contrary to expectations, it did not stop. Tinkler had been handed a life raft, but instead was swimming towards the storm.

8.

JUDGEMENT DAYS
LET'S BURN A DEBT

The year 2012 had started badly, and it soon went from bad to worse. On 26 January, Australia Day, Tinkler learned that his coal loader proposal had finally been rejected by New South Wales premier Barry O'Farrell.

Having lobbied so hard, Tinkler was ropable. It was Newcastle's loss: the residents had 'clearly voted for change and are now being asked to accept that coal will forever be railed through the middle of Newcastle suburbs and the townships of the Hunter Valley'. It was a terminal blow for Hunter Ports, whose logo would remain on the Newcastle Knights' jerseys for two more years, almost as a reproach.

Once the loader proposal failed, Tinkler tried to pull out of his 2011 agreement to pay $17 million for the Mayfield site at Steel River. Mirvac wouldn't have a bar of it. It had a valid contract, and the site wouldn't fetch anywhere near as much from another buyer. Of course it insisted on completion; without any fanfare, its lawyers marched off to the Supreme Court. Did Tinkler think the law only applied to other people?

Tinkler had long believed he was hated in Sydney, and the feeling was mutual. When the loader proposal failed, coming after the break-in, Tinkler decided Newcastle must have it in for him too. Given that he was off the board of the soon-to-be-merged

Whitehaven–Aston, Tinkler was free to live where he chose, and began planning to move overseas. His pad in Maui was kept secret.

After it spent a year on the market, in May Tinkler finally sold his home at Berner Street, Merewether, to his right-hand man, Troy Palmer. The $2.8-million sale sparked a few headlines about the $1.7 million hit Tinkler had taken since buying the property from Andrew Johns four years earlier. Several agents were shocked at the 34 per cent price drop; one tweeted: 'It turns out Nathan Tinkler's sale was a "friendly deal", so should be seen as out of line by valuers.'

In June, the *Sydney Morning Herald*'s Stuart Washington revealed that Tinkler had moved to Singapore; Rebecca and the four kids would soon be leaving Newcastle to join him. It was a defining moment: the 'boganaire' was leaving Australia behind. Tinkler's spokesman, Tim Allerton, said his boss remained 100 per cent committed to his Australian operations, including the Knights and Patinack, but wanted to be closer to the markets in Asia. 'He's got some financiers and investors in Singapore – he just feels it's more comfortable, he enjoys the city and wants to base himself there,' Allerton said. The move had been planned for months, it turned out. The first of Tinkler's Singaporean companies, Bentley Resources, was registered in February.

Singapore was, of course, home to both Ray Zage of Farallon and Will Randall of Noble, Tinkler's two key backers over the years. But most commentators focused on Singapore's business-friendly environment: income tax was capped at 20 per cent, company tax at 17 per cent, and there was no capital gains, death, estate or inheritance taxes. Not that the move would allow Tinkler to avoid tax on his Australian companies, or any capital gains tax on the sale of his Australian assets. But Singapore, often dubbed the 'Switzerland of Asia', has the most millionaire households per capita – 17 per cent – of any nation on Earth, strict laws around client confidentiality in finance, and a tightly controlled media, which, as one journalist noted, 'lets big business get about their business'. For Tinkler, this meant 'no more stories about his house being burgled, his property losses or his dud investment in a supercar club'.

Initially, there was speculation that Tinkler had bought in the expat enclave Sentosa Cove – the only part of Singapore where foreigners are allowed to own land, and where fellow Australian billionaire Gina Rinehart was about to pay $44 million for two off-the-plan units. It turned out he was actually renting, taking up two brand-new adjacent homes in a cul-de-sac at the end of leafy Swettenham Road – Singapore's most exclusive district, once the haunt of Britain's colonialists, where the palatial homes rent for US$25,000 a month. Entered via a long driveway, Tinkler's compound had a water feature and lap pool, and was tended by servants. His neighbours included the actor Jet Li. When the *Courier-Mail* sent a reporter over, they found a Porsche in Tinkler's driveway and a Maserati next door. The twin homes would later spark fresh speculation that Tinkler's marriage was in trouble, although others said the spare house was occupied by his private security guards – unusual in heavily policed Singapore.

*

Even before the Whitehaven–Aston merger was voted up in mid-April 2012, there was speculation about Tinkler's next coal play. Most of the focus was offshore: he was rumoured to be looking in Alabama, Mozambique and Mongolia. But as early as January Tinkler was also having discussions with various junior coal companies in Queensland, including Endocoal. He was offering to invest $17 million and join the Endocoal board, but they chose to get into bed with AMCI and Macquarie instead. Tinkler would try to come back for them later.

A week or so before shareholders were due to vote on the Whitehaven–Aston deal, Tinkler had coffee in Brisbane with Todd Harrington, chief executive of the ASX-listed Blackwood Corporation, which was half-owned by the Noble Group and had exploration projects in Queensland's Galilee Basin. Tinkler and Harrington did a handshake deal: Tinkler would kick in $28 million to take about a third of the company at 30 cents a share, and would join the board. In late April Tinkler set up a new shelf company, Mulsanne Resources – named after the Le Mans racetrack's straight – which he used to invest in Blackwood.

Blackwood shares were trading at only 20 cents, but Tinkler as usual was happy to pay a premium to get his foot on an asset. When the deal was announced in early May, Tinkler told the *AFR* that Blackwood shares were grossly undervalued and he looked forward to joining Noble – with whom he had a 'solid relationship' – on the company's share register. 'Blackwood looks likely to be the first Galilee Basin producer in the marketplace,' said Tinkler, pointing to the 'significant upside'.

The Galilee Basin is a huge thermal coal resource north of the Bowen Basin in central and northern Queensland, and has been targeted by the likes of Gina Rinehart and Linc Energy, who have successfully sold down projects to Indian power companies GVK and Adani for billions of dollars. Galilee coal is generally of lower quality, and hundreds of kilometres of rail and new port capacity have to be built to get it to market. For that reason, it is regarded as a stranded resource with thermal coal prices below US$100 a tonne, but the planning is progressing steadily on the expectation that prices will eventually lift.

Tinkler's investment speaks volumes about his confidence in coal – he was once described as a 'super-bull' – which was so high in May 2012 that he was willing to pay a premium to own shares in an early-stage explorer like Blackwood. It would later emerge that Tinkler's next grand plan after Boardwalk was to roll up the like-minded juniors with interests in the Galilee Basin and on-sell them to a big player, probably Indian, which could then bear the multibillion-dollar costs of building new rail and port infrastructure.

Whitehaven's boss, Tony Haggarty, wasn't standing still either. Within days of the merger, which gave it just under 20 per cent of Coalworks, Whitehaven made a bid for the rest of it. Coalworks owned not just the controversial Ferndale project in New South Wales but also Vickery South, right next to Whitehaven's own Vickery mine in the Hunter Valley. It was a logical fit.

In the *AFR*, Matthew Stevens wrote that the Blackwood and Coalworks deals showed 'the enduring confidence our carbon-hunting entrepreneurs have in their commodity's enriching future'. Stevens noted that Tinkler's bid was at a steep 50 per cent premium to Blackwood's share price, while Haggarty's low-ball bid was at a 'miserly' 17 per cent above the Coalworks price (he would later be forced to sweeten the deal to get it over the line). Price aside, Stevens wrote, both men remained 'true believers', and they weren't alone:

> … a certain class of investor has yet un-shaken confidence in the long story that is Asian thermal coal demand and the downshift in the quality profile of the coal that will fill that demand as India, in particular, responds to the need to meet the electricity demand triggered by its rising middle classes …

Blackwood's and Coalworks' shares rallied on the takeover bids, but the market punished Whitehaven, whose shares had risen slightly to hit $5.58 just after the Aston merger was approved by shareholders, but had then plunged to $4.60 eight days later. Alarmingly, Whitehaven kept falling; soon there were reports that short-sellers (typically, hedge funds betting that a company's shares would fall) were targeting the company. Short-sellers had taken 4.7 per cent of Whitehaven's shares, the *AFR* reported on 11 May, targeting Tinkler, who was 'known to have loans over his assets ... the theory is that some hedge funds are testing how far the stock can fall before any loans are called'. Any exhilaration that Tinkler may have felt after pulling off a third 'deal of a lifetime' didn't last long.

Every time the Whitehaven share price fell a dollar, it cost Tinkler more than $200 million. By June the company was trading at $4. On paper, Tinkler had lost $340 million in less than two months since Whitehaven had been at its post-merger peak. Those paper losses wouldn't have mattered if he could ride out the storm, but Tinkler had borrowed hugely against his Whitehaven stake, just as he had with Aston a year earlier.

In an effort to stem the tide, Tinkler embarked on a harebrained scheme to privatise Whitehaven: he would mount a friendly takeover, and then delist it from the stock exchange. That would stop the share price falling! Tinkler had no funding in place or even the 'highly confident' bank letters that would evidence likely funding. But UBS was in his corner, and Tinkler also had the support of Farallon, which spoke for another 15 per cent of Whitehaven, after he flew to see the company's head honchos in San Francisco. Whitehaven confirmed the indicative bid on 13 June and set up a board committee to consider it. Its shares rose briefly on the news but the rally didn't last: nobody could tell if Tinkler was serious or not.

A month later, Whitehaven confirmed it had received a fuller privatisation proposal from Tinkler, priced at $5.20 a share – a huge premium, given that the company was now trading back below $4. Tinkler claimed to have the support of 48 per cent of Whitehaven's shareholders, who would roll their shares into the bid. The backers were not identified but certainly included Farallon, the Kuoks and Sitorus, and probably also the key shareholders AMCI and First Reserve. Whitehaven allowed Tinkler four weeks to do due diligence.

Somehow, Tinkler now had to come up with between $1.6 billion and $2.4 billion to buy out Whitehaven's minority shareholders. He had conditional support from UBS, JPMorgan and Barclays, but that was not enough to get the bid away. Whitehaven's shares continued to head south; the market was becoming increasingly sceptical that Tinkler would pull off his bid. It's possible that a good number of Whitehaven shareholders were open to the idea of selling their shares for more than they were worth, but quite why anybody would want to join Tinkler in overpaying for Whitehaven was less clear.

It was all an unwelcome distraction for Vaile and Haggarty, who were trying to bed down Whitehaven's merger with Aston and Boardwalk, wrap up the Coalworks takeover, get Narrabri up and running and push ahead with Maules Creek. Any remaining friendliness between Tinkler and Whitehaven's board quickly evaporated.

Tinkler's behaviour was so wayward that for a while it masked the fact that Whitehaven was in trouble. In mid-June the price of thermal coal fell below US$90 a tonne for the first time since 2010, amid growing concern about global economic weakness and the Euro zone's debt crisis. It wasn't obvious yet, but the softness in coal prices would prove more than a market blip.

It is impossible to be definitive about the reasons for the slump in 2012, or to say whether it was cyclical or structural or a combination of both, but there were a few key factors. Firstly, the massive jump in coal prices over the course of the previous decade – peaking in 2008, plunging briefly in 2009 and then climbing again to historically high levels in 2011 – spurred an increase in global coal supply, as new mines were developed and came into production to feed the Asian markets. Higher levels of supply mean lower prices.

Secondly, there was rising concern about future demand for coal amid worsening global warming, possible eventual international action on climate, and likely increased competition from clean energy sources, including renewables and (particularly in China) nuclear power. China's 12th five-year plan included aggressive targets to reduce coal use in the present decade, partly due to climate worries and partly to cut air pollution. The big miners say China has made similar pronouncements before, and has always missed its targets. China watchers say this time the nation is serious, and its coal demand will peak before 2020.

But most analysts, if they had to point to a single cause for the recent fall in coal prices, would probably point to a third factor: the glut of cheap shale gas in the United States caused by the boom in fracking, which has caused many North American power plants to switch from coal to gas. With a smaller domestic market, US coal miners have begun selling their product into Asian markets. Just as China, the world's largest coal miner, had never been a net importer until the end of the last decade, the United States had never been a major coal exporter until this decade. The arrival of a permanent new supplier has depressed world coal prices, with a particular impact on Australia, which is heavily dependent on the seaborne trade into Asia.

These three factors – rising supply, climate worries and cheap gas – weigh heaviest on thermal coal, which accounts for almost 90 per cent of the tonnes traded worldwide. There is work going on to find substitutes for coke in blast furnaces, but it is early days and top-quality hard coking coal isn't found everywhere. Queensland's Bowen Basin is the world's best coking coal resource, and for now it is possible to be reasonably confident that demand from Asia's steel mills will keep rising, even as the 'metal intensity' of development in countries like China progressively winds back. While the outlook for higher-priced coking coal is stronger than for lower-priced thermal coal, the two markets are not completely isolated from each other. There is a cascading effect: falling prices can force lower-quality coking coal to be sold as higher-quality thermal coal, which in turn puts downward pressure on thermal coal prices.

Whitehaven gets roughly two-thirds of its revenue from thermal coal sales, and most of the rest from PCI, a lower-quality coking coal. It was right in the firing line. Its profitability depended on it fetching a solid premium for the PCI it produced, and in 2012 that evaporated. Exacerbating the problem was the high Australian dollar, which jumped above parity on average during 2011–12, cutting revenue from coal sales, which are denominated in US dollars.

What's more, with all their focus on doing the Boardwalk transaction, Tinkler, Vaile and the rest of the Aston team had neglected to do thorough due diligence on Whitehaven itself. Whitehaven's new underground mine at Narrabri came into production in June 2012, and the ramp-up was beset with technical problems.

Narrabri was the key to doubling Whitehaven's production growth from under 5 to over 10 million tonnes a year. The coal quality at Narrabri was a major problem; along with the interminable delay of the approval of Maules Creek, this undermined investor confidence in the company. Some believe Whitehaven knew about Narrabri's coal quality issues as early as 2011, which was why the board tried to put the company on the market. All this had been overlooked by Aston, because Tinkler's personal finances dictated that he do a quick deal with Whitehaven. He hadn't bothered to really examine what he was merging Aston and Boardwalk into.

After releasing some very so-so profit figures in late August, Whitehaven confirmed that Tinkler had dropped his improbable bid to privatise the company. The company's share price tanked again, falling below $3.50. All this dented Tinkler's credibility severely: for many in the market it was a turning point, the beginning of the end.

The fiasco also caught the attention of ASIC, which queried Tinkler over the privatisation bid. Concerned about truth in takeovers, ASIC was considering a proposal to adopt UK-style 'put up or shut up' laws, designed to stop people manipulating share prices by floating merely speculative bids. But if ASIC did in fact investigate Tinkler, nothing came of it.

That same week in late August, federal mining minister Martin Ferguson let slip in a morning radio interview that 'you've got to understand, the resources boom is over'. Surprisingly, for an offhand comment, it got a lot of attention. A bit of psychological adjustment loomed for Tinkler, as for the nation.

*

While Tinkler was struggling to raise the money to buy Whitehaven, Blackwood held a shareholder meeting to approve the $28-million share placement that would give Mulsanne Resources a 34 per cent stake. Blackwood's shareholders duly approved the transaction on 12 July, and the company had expected its money within a week. But on 1 August Blackwood reported that it had given Mulsanne a fortnight's extension. It wasn't a great look: if Tinkler couldn't fund a tiny share placement, how could he possibly fund his $5-billion privatisation bid?

On the very same day, a New South Wales Supreme Court judge ruled that Tinkler's Buildev subsidiary Ocean Street Holdings would indeed have to complete the purchase of the Mayfield site it had agreed to a year earlier, and ordered it to pay Mirvac the full amount owed – $17 million, including interest – by the end of the month.

From this time onwards, a little band of journalists was kept busy running from one Tinkler-related case to another. Covering

Tinkler was almost a full-time job: there were bushfires everywhere. From matter to matter, adjournment to adjournment, Tinkler's snooty legal team appeared to have just one overriding goal: delay. It seemed an abuse of process, wasting court time to the detriment of Tinkler's creditors, who were forced to hold on through interlocutory proceedings, in some cases bearing mounting lawyers fees, without substantial grounds for the delays ever being presented.

This was simply an extension of the old Tinkler modus operandi – use creditors to provide interest-free loans – but now it was being aided by the court system. Judge after judge, registrar after registrar allowed the circus to carry on. Invariably, when the patience of a court had finally been exhausted and a non-negotiable deadline loomed, disputes would be settled, but it would be right at the last minute, sometimes even on the courtroom steps, generally including full payment but on confidential terms.

Mirvac's dispute with Tinkler was a case in point. The merits of Buildev's argument, such as they were, had already been canvassed in the Supreme Court between May and August 2012, by which time an order to complete had been made. On 10 September, after Buildev had failed to comply with the order and Mirvac had sought to enforce the prior ruling, Tinkler's lawyers served notice that they wanted to file a defence.

Mirvac's spokesman was mystified:

> We are surprised that the Tinkler entities have today requested an extension from the court in order to put on evidence, apparently in relation to their finances. We are uncertain as to what their purpose is, and believe that it runs counter to assurances provided over many months that Tinkler entities were more than capable of completing the contract.

The registrar duly gave Buildev another extension, until Mirvac threatened that if Buildev did not come up with the money, it would appoint a liquidator to take over the company. Still it took another six weeks and four more hearing dates for the ludicrous dispute to play out, although Buildev never had a leg to stand on. On the last day of the case, at the end of October, Tinkler's lawyers were still arguing: they sought and obtained another two-hour adjournment, until Mirvac's barrister finally compelled Buildev to pay up every cent it owed, plus legal costs.

In mid-September it emerged that the contractor Sedgman, which had been engaged to work on the design of Tinkler's coal loader, was owed about $2 million for invoices dating from March to October 2011, and was negotiating with Hunter Ports. Two weeks later it applied in the Queensland Supreme Court to wind up the parent company, Tinkler Group Holdings.

The affidavits show there was never any genuine dispute about the existence or the size of the debt. In fact, a letter in February 2012 from Tinkler's then right-hand man, Troy Palmer, to the managing director of Sedgman had acknowledged the balance owing; he'd also promised that if Hunter Ports could not pay by 30 June – more than four months away! – the Tinkler Group would. Even when dealing with a well-run, listed company like Sedgman, Tinkler just wouldn't pay until he absolutely had to. Where Tinkler would find Blackwood's $28 million remained unclear.

*

Of course, what was motivating Tinkler's courtroom shenanigans was another acute cash-flow crisis. As always, it was the Patinack business that was the biggest drain. By Tinkler's own rough estimate, it was costing him $2 million a month. By mid-2012 he could no longer afford it.

In August the *Sydney Morning Herald*'s Tom Reilly broke another exclusive: two months earlier, Tinkler had offered to sell the whole of Patinack Farm, including its sprawling stud farms in two states and more than 1000 racehorses, broodmares and stallions, to Qatari Sheikh Fahad al-Thani for $200 million. But the sheikh had knocked him back, with one source familiar with the offer telling Reilly it was 'wishful thinking' on Tinkler's behalf.

'Even if the Sheikh wanted to take on such a huge number of horses, $200 million was over the top. Tinkler may own hundreds of horses, but there isn't a lot of quality,' the source said. Like Tinkler, Reilly wrote, Sheikh Fahad was a wealthy newcomer to the racing industry, but unlike Tinkler, he had invested cautiously and patiently and with much more success. Sheikh Fahad's bloodstock adviser, David Redvers, told Reilly: 'People often come to us with offers and we reject nearly all of them.'

What was more, Reilly reported, after the knockback Tinkler had turned for help to his mate Gerry Harvey. Patinack had then announced a special reduction sale of 350 broodmares at Harvey's Magic Millions on the Gold Coast, and Harvey was believed to have lent Tinkler $20 million in cash, as an advance on the sale, which would be held at the end of October. Harvey all but confirmed it, telling Reilly: 'Any deal or business we do is a personal matter between him and me. I wouldn't speak about that publicly.' Later reports had Harvey charging 8 per cent interest, and a 12 per cent commission on the sales.

Next, Reilly found out that Patinack had not been paying superannuation to its workers since November 2011 – the better part of a year. Internal documents showed the company had made a series of failed promises to make up the unpaid super; following complaints from employees, the Australian Taxation Office had begun an investigation. The employees' wage slips recorded that contributions were being made, but no money was reaching the super funds.

In one internal email seen by Reilly, staff were told the delay was due to 'cash flow management', while in another, a member of the accounts team admitted: 'It's very frustrating for everyone – I haven't got mine either!' One staff member said it needed to be resolved soon, 'or there's going to be a mutiny'. An anonymous Patinack worker told Reilly: 'This guy's supposed to be a billionaire, so why can't he pay people what they're due? There's a lot of anger and frustration amongst the staff. There's a sense that the whole Tinkler house of cards could collapse. People are definitely worried we won't ever see this money.'

At the Group One Caulfield Guineas, Tinkler finally had some luck. His star colt All Too Hard, by now three years old, ran down the raging favourite Pierro, trained by Gai Waterhouse, to win by half a length, bagging the million-dollar prizemoney and – at odds of 12–1 – making a few punters very happy. The win was a vindication for Tinkler and trainer John Hawkes, who had kept the horse out of the Golden Slipper, resting him over the winter after he ran second to Pierro in the Group One Inglis Sires Produce Stakes at Randwick. Now, All Too Hard had turned the tables.

Sired by Tinkler's own stallion Casino Prince, All Too Hard was promising to deliver on his owner's dream. The half-brother to Black Caviar was now one of the hottest stallion prospects in the country; when retired from racing, he could underpin the entire Patinack operation. All Too Hard would back up a fortnight later with a stirring second-place run in the Cox Plate.

Off the racetrack, the bad news just kept coming. In mid-October 2012 it was reported that the Hunter Sports Group had not deposited monthly salaries into the accounts of Knights players, and had also fallen behind with super payments and office staff salaries. Players were calling their agents, and their agents were calling journos.

Hunter Sports Group spokesman Richard Fisk denied there was a problem but accepted 'there has been tardiness in the past ... all pays for Jets, Knights and HSG employees have been processed and will clear today in the normal course of business'. Fisk texted the *Daily Telegraph*: *We are probably waiting on All Too Hards prize money from the weekend ... lol.* But nobody was laughing. Knights captain Kurt Gidley and star hooker Danny Buderus were caught up in the fuss. It was not even clear whether coach Wayne Bennett had been paid.

Bennett told the *Daily Telegraph* he was not worried at all: 'I'm in contact with Nathan on a regular basis. He knows what he's doing and I trust him. Everything he's promised has been delivered to us.' But Bennett had a few sharp words about the impact of late payment on the younger players, some of whom were living on 50 or 60 grand a year. 'Pay day is important to them. Lots of people live from week to week, and in our case it's month to month. I can understand if the money's not in their account, they've got a problem.'

It wasn't just Tinkler's companies that were teetering: he too was personally strapped. A writ filed in the Northern Territory's Supreme Court by Tabcorp's online betting division, Luxbet, sought $136,189 from Tinkler in unpaid gambling debts. The company's claim, which outlined Tinkler's betting records over three years, made fascinating reading.

He had opened up a $250,000 credit betting facility in early 2009, but kept bumping up against the limit. He lifted it to half a million dollars in May 2010, but thought better of it three days later, reversing the change. He had a few big wins and, like most punters, plenty of losses. After a long losing streak he had to top the account up, and by the end of 2010 he'd sunk $805,753 into it over two years. In 2011, the top-ups ceased. By the end of March that year he owed $179,200, and Luxbet wanted its money. Tinkler kept on betting regardless.

Luxbet sent Tinkler dozens of emails, texts and letters, and finally formal demands for payment, and also sent emails and texts to Troy Palmer. In August 2012 the company finally revoked Tinkler's betting facility. Two weeks later Tinkler paid $59,200 but it wasn't enough to stave off the court action. He failed to show up to court in Darwin, and in November orders were made against him for the full amount owing, including interest and costs. Then, at last, he paid.

*

As the number of cases against Tinkler mounted, one in particular was proving a headache. Blackwood still wanted its $28 million, and Mulsanne – a $2 shelf company with no revenue and no assets – didn't have it. At a corporate level, it was that simple.

The most interesting thing about the Mulsanne story was the unravelling of Tinkler's relationship with Will Randall, the Australian head of Noble Energy, who had made him rich in the Middlemount deal and had charted his progress ever since. Tinkler would later claim that he had always planned to fund the Blackwood placement by selling Noble his 75 per cent portion of the original Middlemount royalty stream, which he still shared with Matthew Higgins and which was worth some $25 million to $30 million. Tinkler also had grand plans to make more coal investments with Noble – he was still keen on Vale's Integra mine, amongst others. Randall was the only person who could make all this happen, but he was playing hard to get.

A tortuous saga unfolded behind the scenes as Tinkler tried to stall Blackwood's chairman Barry Bolitho, chief executive Todd Harrington and their lawyers. Correspondence later tendered in court showed how frantically Tinkler was trying to raise the money, all the while juggling his bid for Whitehaven and the many other niggling disputes that were crowding around him.

Tinkler's original plan was that Barclays would lend him $50 million, of which approximately $30 million would go to fund the

Blackwood placement; the other $20 million would go to Aston Metals, which would later float. But the loan was far from a done deal – in fact, there wasn't even a term sheet.

Mulsanne was due to pay Blackwood by the close of business on Friday 19 July 2011 – the same week Tinkler's Whitehaven bid was unveiled. Draft ASX announcements, bank account details and completion checklists bounced back and forth between Tinkler and Blackwood. The emails between their respective lawyers became increasingly desperate as the 4.30 p.m. deadline approached and then passed.

For two weeks Tinkler had batted away every request from Blackwood: he needed to talk to Will Randall. Tinkler's stalwart in-house counsel, Aimee Hyde, emailed Tinkler at 5:47 p.m.: 'the lawyers are super nervous about us not completing today … have you been able to speak with Will yet?' Just after seven p.m., Tinkler sent an email to Blackwood's Bolitho and Harrington, cc'd to Will Randall, Hyde and Keen:

> Hi Gents,
>
> How are you? Busy on a few things at the moment and placement is on my mind I just haven't had a chance to speak to Will yet.
>
> Suggest we call off the dogs [Blackwood's lawyers] until Monday when a solution is in place please as I will be able to finalise this over the weekend.
>
> This is a long term relationship and that view needs to be taken here, we only build value working together not pulling each other apart so please get a lead on your lawyers and let me speak with Will and get a solution around this so we can go forward.
>
> Will be in touch post this discussion, thanks

Bolitho, whose calls and emails had been repeatedly ignored over the past week, responded politely that his intention was 'not to hound' Tinkler; he reiterated Blackwood's commitment to Tinkler's participation and stood his lawyers aside, pending further talks. Tinkler replied he was 'currently in New York and with all I have on I have been flat out as you can imagine'. He apologised for the delay and said he was looking to finalise the placement by 11 August.

Then, finally, Tinkler heard back from Will.

Nathan,

Thanks for sms and hope your ok.

We are all struggling with the lower commodity prices and the direct impact on equity prices. It is a pressure cooker. We all need to deal with our individual issues … and the situation is not going to improve until 2nd qtr 2013.

If you need to vent some frustration you can always call me. You need to keep in mind that we have been supportive on your journey to date. Nothing has changed. The outcome has been a win win to date for both parties.

My recent absence was due to taking annual leave with Simone / kids. I needed some time out. My life is simple – my family and noble. Nothing more nothing less.

The BWD [Blackwood] deal was done to allow us to both commence a consolidation process in QLD. That has not changed … It is not fair to link BWD and VALE deal. We offered to provide you additional time on BWD … you personally told the BWD Chairman that you didn't want it. We did … If you need to take this view … then we both have a problem on BWD.

I know your under pressure. I know you have been backed into a corner by the market. Please see us as part of the solution team. We cannot do it alone, we also need to manage our own risk, we also face similar issues, however with other friendly parties I am sure we can find a solution.

Let me know what you want to do.

Safe travels, Will

Two days later came Tinkler's emailed reply:

TINKLER: hi mate … been trying to call can u give me a buzz please. I am in nyc.
RANDALL: Just departing Gobi desert (about to take off). Land at 10.15am into UB [Ulan Bator, in Mongolia]. Will sms/call on arrival. Hope things progressing.
TINKLER: ok speak then … we will get there … its what we do
RANDALL: Spot on!

But it wasn't spot on. While in New York, Tinkler had met with Barclays' banker, Dave Ellis, on 23 July – four days after the funds had been due to Blackwood – and their discussions were still completely exploratory. Afterwards, Tinkler emailed Ellis directly: 'Thanks for lunch today good to catch up again. Can't believe their wasn't alcohol involved but I am sure we will catch up on that next time.' He attached a presentation on Blackwood and promised a draft term sheet for the loan and some information on Aston Metals, as well as an email introduction to Harrington, who was 'doing a good job as a first time CEO. Has the fire in the belly too and the ASX hasn't ruined him yet.' A draft term sheet was drawn up for Barclays, but they turned him down.

The emails now became tetchier. Bolitho was polite but firm, emailing Tinkler (and cc'ing everyone else) that he needed to meet urgently and would fly to Brisbane at once. He received no reply. The date Tinkler had proposed for finalising the placement, 11 August, turned out to be a Saturday, so the deadline was pushed over to the Monday. Blackwood's lawyers sent Aimee Hyde a draft variation deed and announcements for approval, and she passed them to Tinkler. He replied from his iPhone:

> Hi Aimee,
> I am not signing this until I hear from Will.
> Sick of solving everyone else's problems and being left alone with my own.
> When I get my paperwork off noble I will sign the deed.

On the Sunday before the second deadline, with $28 million due the next day, Tinkler emailed Randall:

> hi mate,
> I am back in Singapore let me know if you are around for a beer this afternoon. I say beer because a coffee just won't do.
> Somewhere around MBS [Marina Bay Sands], would be good if your around
> Trust the family is well.
> Cheers
> Tink

Randall fobbed him off; he was too busy:

> Nathan,
> Thanks email and hope your hanging in your end. Solid floor starting form on coal front, however we all need a few more wins on steel mill side of equation. I have been in HK for 2 weeks and not back until Friday next week into Singapore. We have results call Monday night and then in Indonesia for balance of week.

Suggest we have a call tonight at some point.
Touch base soon.
Will

Whether or not they spoke that evening, Tinkler's next email to Blackwood was furious. It was almost 11 a.m. on the morning of settlement day, 13 August, and he still did not have the money. 'I love how I'm the only one on the hook to honour what they said they will do?' he spat, by now also cc'ing everyone. 'I will do that but it is just taking longer in this market than I would like. I will call you this afternoon Barry.'

Tinkler was still scrambling to raise the funds he needed. He tried Mark Carnegie. He tried the US investment bank, Jefferies. Then he tried Ray Zage, hoping in vain that Farallon might be persuaded to lend him even more money, to get Blackwood off his back. The meeting notes show Tinkler's pitch on Blackwood: 'I am keen to acquire other Australian assets with this vehicle such as Endocoal and Integra ... This will be a strong company and will have a market capitalisation of $1bn within 12 months. This is a very similar event to the Boardwalk story.'

Zage declined. Tinkler was now navigating a treacherous passage between Noble and Farallon, either of which might sink him. So another settlement day passed without result.

In late August Bolitho lost patience and gave Tinkler a week to show how he was going to fund the placement, or Blackwood would issue a formal statutory demand. Tinkler did nothing, and on 31 August the demand was issued. If Mulsanne didn't pay within 21 days, Blackwood would wind it up.

Tinkler again tried to reach Will Randall, who emailed back that he would be in Singapore the next morning coaching his son's rugby team, and proposed a catch-up. That night, Randall emailed the Blackwood board: 'Trying to keep it polite. Welcomed him into my home today for coffee. No response via email or phone. I didn't chase him either. He has no solution and talk is cheap. Sorry team.'

Tinkler and Randall did catch up the next day. On the Sunday evening, Randall again emailed the Blackwood board: 'NT and I met at kids rugby this morning and it will be a weekly event until 20th December. Both playing at the same club.' Tinkler was primarily focused on finding a solution for his Whitehaven merger, Randall said, and his disputes with Mirvac and Blackwood were secondary. Ominously, Randall concluded: 'in short and my humble opinion – no solution will be forthcoming until full legal due process is fully exhausted'.

For his part, Tinkler was trying another investment banker – Credit Suisse – minus his usual charm. He attached a draft term sheet to fund the Blackwood placement with a $30-million loan, describing it as 'a quick one to get in our deal book' which had a 'lot of short term upside'; given that Credit Suisse had a track record of backing junior mining companies, this would be a 'nice one to get away which I think we can tie into other things later'.

When the bank came back with a few questions, Tinkler fired back:

> Pretty obvious you didn't look at any of the info that I sent you. All these questions were answered ... this is an exploration company which is in feasibility stage. A lot of corporate activity can be brought to play here. I hope you can be involved, will be very lucrative for CS.

But Credit Suisse's qualms were well founded. Blackwood had no feasible mining plans or even resources as yet, and the more the compliance team dug, the more bad publicity they found. Blackwood's recent announcements about pending legal action over the placement did not bode well, and the bank's internal correspondence showed that it was worried about it: '[Tinkler] would need to update us on current status and why he is still looking for financing regarding a share placement that, according to stock exchange announcements, is definitely off.' Credit Suisse was a no.

With four days to go, Tinkler emailed Hyde: 'Please draft a letter to Blackwood we don't want this going any further.' He would seek another month's extension. Blackwood refused, unless Mulsanne could provide proof of funding the next day. Mulsanne couldn't, of course, and on Monday 24 September, two months and three deadlines after the share placement was due, Blackwood began wind-up proceedings and made an announcement to the ASX that morning.

Tinkler snapped, sending an explosive email to Bolitho, Harrington, Randall, Elman and others:

> Hi Gents
>
> The below is just a complete act of bastardry. Have made many attempts at this over the last 2 months but with the adverse press etc I have been copping that has been extremely difficult to put in place and I have discussed this with Will and corresponded with your lawyer at this time. What was to be lost by giving me 30 days to announce another deal I am working on and this would fall into place. Now for the first time this placement is in jeopardy.
>
> Tough to have any sort of relationship with any of you based on these actions going forward. Was in touch with your lawyers several times last week and we were trying to finalise a time extension but there was never any mention of this as a path forward.
>
> Disappointed in all of you particularly those I have known or a long time. Enjoy the headlines you have created tomorrow. I am sure reading them you will all feel vindicated, make

sure to get a quote in Fairfax will be in touch with each of you I am sure.

Thanks for making things even more difficult at what is already a hard time.

Nathan

'Well it seems we have finally got his attention,' wrote one Blackwood director, Andrew Simpson. Bolitho chimed in: 'Both Todd and I have had voice messages this afternoon from NT with language a tad stronger than his email. As we are aware we have gone to great lengths to engage him but with no response.' But it was Randall who put the nail in the coffin:

Gents

Emotion has no place in business and please let it go thru to the keeper. Not worth it. Similar this end of the world. Also not the first time.

We have a process to follow for all shareholders (large and small) and the BOD [board of directors] is executing accordingly.

An hour or so later Tinkler, fat-fingered with fury, emailed Zage from his iPhone:

Hi Ray,

You new to e aware of the below it is going I be more good press for me. I have met with Will several times over the last few weeks to extend timing and have been assured it is fine but they have gone ahead and done this.

I have grounds for defence but Mulsanne is a $2 company do can be wound up if necessary and they no that but have chosen to be dramatic regardless.

Just another one lining me up. Happy to get Aimee to talk through with Ashish so you are comfortable if need be but it does not effect any structure of ours as I am sure you were aware. I do have funding lines up but no doubt that will disappear now for a third time following the press.

Nathan

*

Just as Tinkler's corporate antagonists were lining up to sue him, a new front opened up, one that would arguably do more damage to his reputation than any of the barneys playing out in the business pages.

Right from the beginning, as Tinkler scrambled to the top, he had left a trail of unpaid creditors behind him, but it wasn't yet widely known. The personal disputes that had broken out in the public arena – with Anthony Cummings, with Tim Sommers, even with Branko Culina – were somehow seen as going with the territory. As Tinkler's friend John Singleton told one newspaper: 'You don't get to where Nathan's got to without treading on a few people.' The public had perhaps recognised that, as Tinkler embraced a rich man's world of horseracing, supercars and football teams, he might easily become a media target.

By 2012, however, the *Newcastle Herald*'s journos were starting to cop flak for all the glowing coverage Tinkler was getting. Hadn't they heard the stories? They started digging, and it became clear that, below the radar, completely unreported, there were literally dozens of 'little people' who had been strung out by one or another of Tinkler's operations. Trade suppliers to his stud farm, subcontractors on his building sites, consultants for the Hunter Sports Group – they were everywhere, and their stories were often sadly alike. Late payment. No payment. Bullying or intimidatory tactics. Rude, downright offensive treatment.

The problem was getting anyone in Newcastle to speak out about it. Some were still hoping to get paid and knew that if they went public, they would go to the back of the queue. Some were simply intimidated, fearing they would be taken to court if they spoke up – and how could they ever hope to come off best against a billionaire? Others just wanted to leave it all behind them. To this day, a mix of fear and loyalty stops many Newcastle locals from talking about Tinkler.

The dam wall burst on a Tuesday, 21 August 2012, when the *Newcastle Herald* covered its front page with a photo of a smiling Tinkler, dressed in a suit. The headline read 'Trail of debt: Hunter firms call on Tinkler to pay the bills'. The story, by *Herald* reporter Donna Page, opened: 'He is estimated to be worth $915 million and is said to be one of Australia's richest men, but Nathan Tinkler has left a trail of debt devastating small-business owners from the Upper Hunter to Queensland.'

Page had spoken to dozens of businesses in the Hunter, 'ordinary people, mum-and-dad businesses, many individually owed close to $100,000. Some have been forced to remortgage their homes, increase overdrafts or sell their cars because they have not been paid for work or goods supplied.' An unnamed spokesman from the Tinkler Group seemed to acknowledge the problem, admitting that creditors were paid out on 'an ongoing basis'. Most debts were owed by either Patinack or Bolkm, the construction arm of Buildev, which Tinkler controlled.

There was the Newcastle building industry veteran Bob Jeffkins, who had employed a debt collector to chase $17,000 owed to him by Bolkm:

> I am just glad I got it down to a relatively small amount compared to the figures they did owe us. I have no interest in ever doing business with them again, there are many people owed money and it's just not the right way to do things.

There was Phil Hohl, of the Queensland-based Hohl Plumbing, which got court judgements against Patinack Farm over a debt of $5681, and against Bolkm for more than $29,000. Hohl had successfully recouped more than $170,000 on behalf of other small creditors, and told Page:

> I do not like to see innocent battlers out of pocket when they are rightfully owed money. I can tell you now I wouldn't give Nathan Tinkler credit for a pack of biscuits, I always find it amazing when companies associated with wealthy people don't pay their bills.

There was the 30-year-old Maitland builder Mitch Fraser, who sold everything, including his car, to keep his business afloat but was eventually forced to go into administration after 14 years in the industry. Bolkm was his largest debtor, owing him $33,000 – down from $200,000 at one point. He told Page:

> They gave me about 20 different payment schedules and never stuck to one … They had us on the drip for a long time, we were barely surviving and I was at my wit's end. I was selling everything and doing all kinds of nasty things just to keep going, but I couldn't keep carrying on with what we were owed. I paid everything I could and I will do my absolute best to make sure the two creditors I have left are eventually paid. I have no idea what I am going to do now, I've lost everything, but I can sleep at night knowing I did the best I could to do the right thing. It was very important to me to do that.
>
> This is just the absolute tip of the iceberg, people are too scared to speak out because they are desperately hoping to be paid eventually. It is unbelievable this is happening when you look at Tinkler's high-rolling lifestyle.

There was Alice Russell, who ran Mudgee Chaff and Grain Milling, which refused to supply Patinack Farm after fighting to get a debt of $17,000 reduced to $3500:

> I consider myself one of the lucky ones, because there are people all throughout the valley owed a lot more than me. We always felt sorry for the Patinack employees, it wasn't their fault. But I want nothing more to do with the business.

There was the roofer Rob Atherton, owed $50,000 by Bolkm; he had not paid himself for a month to make sure that his employees' entitlements were up-to-date. 'They just keep telling you … that it will be paid on this day and that day, it really pisses you off,' he said. 'It's been an absolute nightmare and a real struggle to keep our heads above water and keep going.'

A former Tinkler employee told Page:

> They drip feed people in an effort to keep them quiet and keep things ticking over. It's the ones who scream the loudest and threaten to take it further who get paid first. It's a terrible way of doing business. If a business cuts them off, they just move to the next supplier or sub-contractor and it all starts again.

Page revealed that, over the previous 10 months, seven business had succeeded in getting court orders against either Patinack or Bolkm to recoup about $112,000 in debts, owed to companies including Hohl Plumbing, Agricultural Publishers, Mercedes-Benz, Amgrow, Belmark Rural and Cygnus Transports. The individual debts ranged from $38,497 right down to $1295.

The day after her sensational splash, Page reported in the *Herald* that Tinkler had finally started to pay off those small creditors. Bob Jeffkins was in shock:

> Half a dozen people rang me to say they got their money through – if the paper had not done a story it never would have happened. They have strung everybody out to the nth degree and it's made things bloody hard for a lot of people who have suffered a lot of hardship. I just wish we had gone public with this whole mess a long time ago because they obviously don't like the way it looks.

Rob Atherton said he had received $15,000 but was still owed $30,000:

> I just want what they owe me, it's as simple as that. I'm going to be contacting my solicitor about chasing the rest. It's pretty amazing how one minute they don't have the money and the next they do.

Over three weeks, Page estimated, she spoke to 60 businesses owed money by various Tinkler entities.

On 1 September Page reported that medical providers were chasing the Hunter Sports Group over unpaid bills issued to the Jets and the Knights. Some had employed debt collectors. One doctor had refused to see players without payment up-front; another had started billing players directly. Doctors told Page that phone calls and letters of demand to the Hunter Sports Group were routinely ignored. 'We don't have these problems with other patients, it's absolutely unbelievable,' one doctor said. Most medical providers did not want to be named, partly due to concerns about the players' privacy.

The same problems were reported by other team suppliers and contractors, but some had been waiting for six months. One told Page: 'We thought when Tinkler took over the clubs there would be no problems but the situation is getting out of hand.'

Relations between Tinkler and the *Newcastle Herald* had been getting worse all year. On one occasion, Palmer and Fisk went in to heavy the paper's editor, Roger Brock, and Newcastle Newspapers' general manager, Julie Ainsworth, complaining they 'had it in for Nathan'. In the newspaper's own boardroom, they made insinuations that the paper had done nothing for the Knights – forgetting the huge support the paper had given Tinkler's bid, and ignoring years of sponsorship. Newcastle born and bred, and from a rugby league family, Ainsworth lost her cool, telling them where to go and storming straight out.

After Page penned a story on 11 October about a $350,000 tax bill the Knights owed, Tinkler snapped. The Hunter Sports Group, claiming it was the victim of a negative campaign, banned *Newcastle Herald* journalists from attending Knights and Jets press conferences or other media events. Staff and players were banned from talking to the paper's journalists. It got toxic: editors and journalists were hounded and abused. Not only was the honeymoon between the city's highest-profile businessman and its proud local paper over, but a bitter divorce was being played out. Palmer blamed the 'tall poppy syndrome'.

New editor Chad Watson stood firm. The stories kept coming, as more and more people came out of the woodwork, not just in Newcastle but right around the country, to tell of their encounters with Tinkler. From then on, Tinkler would never shake the trail of creditors that followed him.

9.

DOWNFALL
RUN FOR NATHAN

Tinkler was now on the warpath. Whitehaven, still the main game, remained in his sights. But even if he had the money to launch a new, lower takeover bid, he was blocked by a six-month 'standstill agreement' he'd reached in July, a condition of access to the company's books. Thwarted, Tinkler decided to try and roll the Whitehaven board, as he had done before with Coalworks and Aston. Things were about to turn ugly.

The campaign started with a few strategic leaks to the press. At first, it was unclear from the reports whether Tinkler would move against his erstwhile allies Vaile, Christensen and Flynn. Tinkler had already shown that Vaile could be pushed aside. Presumably his next target was Haggarty, whose position was increasingly uncertain – he was on a rolling one-month contract, there were reports he might not stick around, and he refused to confirm or deny anything, which fed speculation that Tinkler might install Flynn as managing director of Whitehaven in his place. When asked about Tinkler's intentions, Vaile prevaricated, telling the *AFR* that any move on the MD was news to him.

A tense three-way standoff was emerging. In one corner was young Tinkler, the largest shareholder but clearly under financial duress. In the other was the older Haggarty, with experience and credibility but half an eye on the exit. Between them was Vaile:

Tinkler's man, no miner, but a savvy political operator looking to chart a way forward and keep his boardroom reputation intact.

The dynamic on Whitehaven's board was fluid. When its annual general meeting notice came out, it revealed that Christensen and Flynn were formally employed by the Tinkler Group until the end of September, but were standing for re-election to the board as independents. Ostensibly Tinkler's representatives, they appeared to have 'gone native'.

A week out from the AGM, a Whitehaven board meeting was interrupted by calls from the *AFR*'s Jamie Freed: had Haggarty seen Tinkler's letter? The newspaper had a copy: Tinkler was demanding that Whitehaven release full financials and confirm its earnings guidance within 48 hours, asking that he be released from his standstill agreement, and threatening to vote against all resolutions put to the AGM.

It was a fundamentally hostile act by the company's largest shareholder. Under the 'two strikes' rule brought into the corporations law in 2011, if Tinkler succeeded in getting a 25 per cent protest vote against Whitehaven's remuneration report, he would register a first strike and be halfway to a spill. It was purely a tactic to embarrass the directors – an actual spill would require a second strike at the next AGM – but no board wants a strike recorded against it. Owning 19 per cent of the shares, and with the directors prohibited from voting, Tinkler could easily get over the line. Whitehaven had no choice but to respond. Tinkler had burned his bridges.

Vaile refused to drop the standstill agreement. He wrote to shareholders, saying the board could 'not understand the motivation of the Tinkler Group demands'. Haggarty refused to put out any new financial information ahead of the scheduled release of the company's quarterly report, which was anyway due only a day after Tinkler's purported deadline.

Sure enough, when Whitehaven's quarterly came out, it made for pretty grim reading, with a tacit profit downgrade blamed on the fall in coal prices: thermal coal had dropped from above US$90 a tonne in July to below US$80 a tonne in October. If those prices persisted, Whitehaven would not make $185 million in 2012–13, as analysts were expecting, but just $50 million. Underscoring the point, Whitehaven had put the Sunnyside mine on care and maintenance, indefinitely. And there was a separate announcement: a succession planning process was underway. For Haggarty, who had already told Vaile he would not stick around if Tinkler's privatisation bid succeeded, the latest debacle had clearly been the last straw.

Tinkler ramped up the pressure. On Wednesday 31 October, the day before the AGM, he placed an ad in the *AFR*, in which he lambasted the board over the 44 per cent decline in Whitehaven's share price during the six months since the end of April – a loss of $2.3 billion in shareholder funds – and declared he would vote against every resolution put to the meeting. The ad came too late to make much difference to the result, given that most institutions had voted by proxy the day before, but still it made for gripping reading.

Although he acknowledged that macroeconomic uncertainty and falling commodity prices were partly responsible for Whitehaven's share price woes, Tinkler nevertheless pressed all the buttons to take his dispute with the company and Haggarty nuclear. He attacked Haggarty's personal decision, announced 35 days prior to a downgrade, to enter into a 'collar' arrangement over 8 million of his own shares with Deutsche Bank – which, in effect, insured him against further volatility in the share price. He attacked 'adverse off-take agreements' and undisclosed 'friendly marketing arrangements'. He attacked the company's progress at Maules Creek and at Narrabri. In short, Tinkler threw every accusation he could, fair and unfair, at the board in general and at Haggarty in particular. He closed with a broad assertion:

> Unfortunately, since the merger with Aston Resources, Whitehaven has had limited development success, torn through nearly half a billion dollars in capital and drawn down previously unextended bank facilities ... [I] have little faith that Whitehaven will deliver any profit this year.

And in case anyone had missed the ad, Tinkler then gave an extraordinary interview to the *AFR*, ensuring that it had a sensational front page the following day. Letting fly from Singapore, Tinkler said that Haggarty 'spent more time with his cows in Gunnedah' than with his management team and that Vaile was 'a politician [who] doesn't know anything about mining'; he also called Christensen a 'piece of gum on the bottom of my shoe'.

For good measure, he accused Whitehaven's directors of fanning media speculation about his personal balance sheet. 'I have no doubt there's been more emails between the Whitehaven board about articles written on me than has been discussions around the problems at Narrabri,' he said.

Whitehaven had been a 'constant source of worry', said Tinkler. 'I can't stand by – I've got far too much invested to stand by and watch this thing continue to fail.'

Without going into his personal finances, Tinkler also squarely addressed the perception that he had fled the country to escape his financial obligations. 'All I will say is that tough people last, tough times don't,' he said. 'There's been a massive media campaign to try and ruin me ... they wanted to treat me like Christopher Skase, like I was sitting in a wheelchair puffing in a bag and owed $250 million – it is just not the case.'

It was vintage Tinkler: personal abuse, directing blame at everyone but himself. It didn't help. If anything, it was conduct unbecoming in business, and it galvanised opposition to him.

Whitehaven's annual general meeting, to be held at the Sofitel

Wentworth in Sydney, was sure to be a doozy. Vaile told reporters he hoped it would be a 'watershed' moment for the company. The sense of occasion was heightened by an anti-coal protest outside the hotel, and by a brief skirmish inside as a dozen or so greenies who opposed the Maules Creek mine sat with their arms linked, attempting to stop shareholders entering the meeting room. Anxious attendants dashed about; within minutes, the protesters had been hauled outside by police.

Slowly the room filled, and soon it was standing room only. There were a couple of hundred people, mainly small shareholders but also plenty of suits – bankers, fundies, analysts, collected hangers-on – from all sides.

Once again, Tinkler himself was nowhere to be seen. Although he had effectively turned the meeting into a referendum on himself, he did not show up and put his case. Whitehaven's board took to the rather small stage. Vaile opened proceedings, his skills as a politician coming to the fore. Here was someone used to drama and unfazed by the public brawling. When it came time to vote, he threw the meeting over to the floor. 'I always enjoyed question time,' he quipped.

Vaile first put up a slide showing the proxies he held. On the crucial resolution of whether to approve the remuneration report, Tinkler had 24.98 per cent of the votes, meaning that even with his 19.4 per cent stake, and with directors prevented from voting, he had fallen just short of a strike.

The Australian Shareholders Association's Stephen Mayne, taking the floor with about 140,000 votes in his hand and ready to vote against the report, switched to declare he would support the board; Tinkler was 'pretty friendless' on the share register, he noted. For Mayne, Tinkler's attack was an abuse of the two strikes rule, given that it had nothing to do with remuneration. As the votes were counted, it became clear there was little support for

Tinkler in the room; several grumpy small shareholders were happy to tell reporters they were voting against him. The final vote showed just 24.78 per cent were opposed to the resolution. Tinkler's campaign had bombed.

Vaile, Flynn and Christensen were all elected by similar margins, despite Tinkler. Ahead of their re-election vote, Mayne took the opportunity to grill Flynn and Christensen, who revealed publicly for the first time that they had gone on six months' 'gardening leave' when Aston Resources had shut its Sydney office ahead of its merger with Whitehaven. Now standing as independent directors, they wished to distance themselves from their former employer.

Mayne wanted to know why they had quit working for Tinkler. Flynn stonewalled: he had known Tinkler for many years, they had 'some different views', there was a 'parting of the ways', it was a 'natural thing'.

Awkwardly, Mayne asked Christensen whether Tinkler had paid out all his entitlements. 'I'd prefer not to answer that as I think the chairman said we could talk to you about that after this meeting but I don't think it's productive here,' Christensen replied.

Mayne also pressed Vaile to defend his relationship with Tinkler. 'Do I feel embarrassment, or silly? No,' said Vaile, rather unconvincingly. Vaile heaped praise upon Haggarty and reassured everyone that the managing director's departure was not imminent, that the succession would be orderly.

Straight after the meeting, Vaile and Haggarty held a joint press conference. Handing out a written statement, an emotional Haggarty said the lack of support for Tinkler's value-destroying attack on Whitehaven was evident from the day's vote. While he was not inclined to dignify Tinkler's vocal attack on the board with a response, it was not appropriate to let the misleading and inaccurate allegations of the past week go unchallenged. After 30 years in the coal business, said Haggarty:

... I believe people should be judged by what they do, not what they say, and I am happy to have the track record and reputation of me and the Whitehaven team compared to that of Tinkler Group at any time. I suggest that Nathan Tinkler focuses on getting his own house in order and leaves the running of Whitehaven to those who have the qualifications and experience to do so.

Vaile, who could not get close enough to Haggarty at this point, made it clear that Tinkler was a lone voice opposing the board: '[In] a quick analysis of the proxy votes that have come in, it also seems that some of those financial institutions that have a shareholding in the company that have supported the Tinkler Group over the years have voted in favour of the board and the remuneration report.'

Vaile was referring primarily to Farallon, and it would soon be confirmed by various journos that the hedge fund had indeed abandoned Tinkler because of his crazy attempt to force a spill of the board. Tinkler had been told to pull his head in, by the one person he would listen to – who, in reality, now had him by the short and curlies – Ray Zage.

*

On the same day he was dumping on Whitehaven to the *AFR*, Tinkler had given an extensive interview to the *Daily Telegraph*'s racing writer Ray Thomas, explaining that he was prepared to sell Patinack Farm at the right price. He told Thomas that running the stud was 'not as enjoyable as it once was'.

I love my racing but I'm living overseas now and I'm not as hands-on as I'd like to be … There is no one in the horse industry that wants to speak positive about anyone. The racing industry is a self-promoter's dream and I am not a self-promoter. If there was one thing that I could change about myself, that would be to like reading about myself and seeing myself in the papers but I just don't.

Tinkler told Thomas that his football clubs were 'safe'. They were community-based, hands-off and were not costing him anything like the amount of money that Patinack did:

Racehorses take management, man-hours, lots of people, all those sort of things. Fortunately, there is no manes on most of those footy players so they can think and speak for themselves. The footy clubs are a much easier business to manage – horses is just massive, the money is massive, everything is just too hard.

Tinkler put his own spin on the *Sydney Morning Herald*'s August report that he had tried to sell Patinack to Qatar's Sheikh Fahad al-Thani. 'I wouldn't sell it to him,' said Tinkler. 'I don't think Australian racing needs more sheiks. Australian racing needs characters and people they can relate to, not more faceless foreign owners.' In the same breath, though, he told Thomas he'd consider selling to or teaming up with a Chinese partner.

Tinkler also denied that the reduction sale at Patinack had anything to do with the fact that his thoroughbred racing and breeding empire was bleeding financially; it was more about getting the number of horses he owned back down:

Earlier this year, one of my staff told me that I was going to have 1800 horses shortly. I just said: 'Mate, what?' I believe a more sustainable level is about 800 to 900 horses. This is a bad time to be selling but we have so many foals hitting the ground we had to do something.

After his defeat at Whitehaven's AGM on the Thursday, Tinkler got a consolation prize at Flemington that weekend, when his three-year-old filly Nechita won the Group One Coolmore Stud Stakes on Derby Day in Melbourne, at odds of $5, winning half a million dollars in prizemoney.

It was a huge win for Nechita's emotional head trainer, John Thompson, who let his guard down the next day on Sky Sports Radio, saying cash-flow problems had affected Patinack 'to an extent you wouldn't believe':

> The last few months has been pretty tough for the guys at Patinack Farm ... I have gone weeks without farriers, bedding, run out of feed a number of times. It's a cash-flow problem. Nathan is obviously very wealthy but it's all paper sort of value. He has always suffered from owning a business that doesn't have that cash flow and it is that cash flow that pays the bills. In the last 12 months, we have just basically been relying on prizemoney.

Thompson also confirmed rumours that Patinack had been kicked out of its Hawkesbury private training facility: 'the owner of the property just got sick of us being late with the rent'.

Thompson wasn't putting the boot in: he remained loyal to his boss, saying Tinkler would 'come through the other end ... I have good faith in him and he is very confident things are going to happen.' But the revelation that Patinack's horses were going without feed – which was subsequently denied by Tinkler's spokesman, in an effort to hose the report down – was extremely damaging. If true, it was verging on animal cruelty.

Tinkler suffered another big hit to his reputation a fortnight later, when the *Telegraph* reported that he was on the verge of losing his private jet. He'd fallen behind in his monthly repayments to financier GE Capital, which had taken out a repossession order and had the Dassault 900, worth $16 million, grounded in Bankstown. On Thursday 15 November 2012, as Mulsanne was nearing liquidation, a pissed-off Tinkler flew to Singapore regardless. A source close to GE told the *Telegraph*: 'Mr Tinkler was instructed by GE to leave the jet at Bankstown but has blatantly ignored all advice by flying the plane to Singapore as much as he feels like.'

GE, owed $12 million, put Tinkler's private company TGHA Aviation into receivership a week later, also grounding his $5-million Augusta helicopter. Both were soon put up for sale. It was a moment laden with symbolism: the private jet and helicopter are the mark of the billionaire. Tinkler tried desperately to get them back, telling the receiver that he hoped to pay out GE. But as receiver Nathan Landrey, of Taylor Woodings, told the *Telegraph* in January 2013: 'I can't stall the process of receivership every time a director floats that they are going to refinance.'

From November onwards, the wind-up applications were coming thick and fast; as soon as one creditor was paid, another would take its place. The chief commission of New South Wales' state revenue moved against Patinack Farm, which owed $259,866, and as soon as that was paid Workcover took over. Another company, Tinkler Group Holdings Administration, owed the state government $82,493; as soon as that was paid, law firm Gilbert & Tobin stepped up, owed roughly $345,000. When that was done, private security company Internet Fraud Watchdog substituted themselves as creditors, owed $145,000. The New South Wales government then sued the Hunter Sports Group for $600,000 in unpaid rent on Newcastle Stadium. Queen Street Capital and Aston Copper also faced a wind-up in Brisbane by the law firm HWL Ebsworth. Tinkler had failed to complete on the purchase of 443 Queen Street, and the real-estate agent McGees Property was suing for its fee.

It was tough just keeping track of it all. In late November 2012, just days after the court approved the wind-up of Mulsanne, a second Tinkler entity fell into liquidation, and this time it wasn't a shelf company. At a Federal Court hearing in Adelaide, Workcover South Australia succeeded in winding up Patinack Farm Administration over its relatively measly debt of $16,798. Tinkler's people explained that the missed payment was due to administrative error; that was easy to believe, given the collapse of the business empire that appeared to be occurring around them.

Patinack Farm Administration held no major assets but was the main employer at the stud, responsible for almost 200 workers, many of whom were owed unpaid superannuation. The liquidators issued a statement calling for employees to contact them and detail their outstanding entitlements and other claims.

Would the employees still turn up to work? Would they be paid? Once the wind-up had begun, it was all a mad scramble; awkwardly for Tinkler, even though the tiny debt was paid off as soon as it was discovered, there was no provision for the court to simply set the process aside.

Standing behind many of these wind-up applications was the most persistent creditor of all: the Australian Taxation Office. On Wednesday 13 December 2012 came another bombshell: the ATO had moved to wind up the Newcastle Knights and the Newcastle Jets, along with their parent company, the Hunter Sports Group, over unpaid debts of some $2.7 million.

Tinkler's spokesman was genuinely shocked and, recognising the gravity of the situation, issued an immediate statement: 'We are surprised by the move by the ATO, as we have not received notification of this move … we advise that any outstanding sum will be paid as soon as possible.' The NRL said that it was monitoring the situation but had received 'assurances from the Knights that the club is financially stable'.

The electronic media went into overdrive. Threats to the survival of Nathan Tinkler's business empire were one thing; threats to the survival of Newcastle's two beloved football clubs were another. Suddenly, Tinkler's collapse mattered.

*

Amid the mayhem, Tinkler also faced a festering dispute with his oldest partner, Matthew Higgins, seemingly over nothing much. Construction of the Middlemount mine had fallen years behind schedule – it had been one of Tinkler's main criticisms of the Macarthur Coal board that had bought the project from him – but finally, after delays due to repeated flooding, the first coal was mined from Middlemount in June 2011. For Tinkler and Higgins, that meant some handy royalty payments under the $1-per-tonne agreement held by Oceltip Pty Ltd, which they owned 75/25. Out of the blue, after a few years of very little contact, Tinkler sent an email to Higgins that July:

Hi Matt

How are you? I trust you and the family are well.

I believe we received 90k into Oceltip from Macarthur this week. I have copied my CFO Michael Wotherspoon on this email so you can liaise with him as to where you would like your funds directed. I believe Middlemount production is due to ramp up before the end of the year, at last.

Let me know if you want to come to a Knights game for a beer sometime.

All the best.

Tinkler signed off as chairman of the Tinkler Group. A few days later, Higgins wrote back:

Hi Nathan,

The family is growing up, way to quickly! I'm sure you feel the same about yours. I bumped into Rebecca and Wilson a while ago and I was shocked how much older Wilson was.

I was up at Middlemount last Christmas between floods, it sounded as though this was going to be a big year for the operations, I was thinking of heading up there for the first train! They'll probably organise it for dog watch!

I wouldn't mind catching up, just somewhere discreet, Brisvegas is probably best.

Anyway finally the 'Middlemount Dream' is sparking into life!

Take care,

Regards,

Matthew Higgins

Higgins tried to follow up a couple of months later, calling Tinkler's mobile. The call was returned by Mike Wotherspoon. Higgins was given a copy of the Oceltip accounts for 2010–11 and suggested that, with money starting to flow to the company, it might be a good time to do some housekeeping: Oceltip had not met in almost four years, nor had it ever had a shareholders' agreement.

A few weeks later, Higgins' lawyer sent him a copy of an ASIC notice appointing Tinkler Group's in-house counsel, Aimee Hyde, as company secretary for Oceltip. As one of only two Oceltip directors, Higgins was miffed: who was Hyde, and why hadn't he been consulted?

When he got on to Tinkler a few days later, their conversation can't have been very friendly. Higgins objected to Hyde's appointment, and Tinkler told him to deal with Wotherspoon. Higgins then emailed a draft resolution to appoint himself as company secretary. Tinkler's response, copied to Wotherspoon, was surprisingly hostile:

I won't be signing that so you can generate as many resolutions as u like. You don't negotiate by yourself and this was never discussed. The tail doesn't wag the dog ...

I am fine with the way things are, my team are doing everything they have to do under the corporations act in relation to your rights as a director and shareholder and the requested information will be provided to you but by all means go ahead and waste your money on lawyers.

You run your accountants agenda but nothing is changing. Looks like you haven't either, not the bloke I knew. Now that MCC [Macarthur Coal] has been bought by Peabody they may be interested in buying this out. I hope so, be nice to have complete separation.

I suggest if that interests you mike writes to MCC and GCL [Gloucester Coal] offering the royalty. I think a price of [price censored] may be achievable. Mike please do that for our shareholding in Oceltip and let matt no the result. If he decides not to sell we still can don't need the hassle. Very sad Matt.

Life is too short.

Tinkler and Higgins had already fallen out once, during the original sale of Custom Mining, so in a sense this was a renewal of old hostilities. While he was offended by Tinkler's personal attack, though, he was intrigued by the price Tinkler was flagging, and no doubt aware that he would need to work with Tinkler to get it. (While Higgins' affidavit blacked out the price, documents filed later in the Mulsanne dispute revealed it was between $25 million and $30 million.)

Higgins emailed back that there should be a meeting of Oceltip directors to discuss housekeeping, a possible sale and how the value was calculated. 'I look forward to an unemotional response,'

Higgins chided. Meanwhile, he wrote to ASIC opposing Hyde's appointment, and had his lawyer inform Tinkler of this. Hyde reversed the appointment and explained it was an administrative move intended to relieve Tinkler of secretarial duties across a hundred companies. She also told Higgins that Tinkler was resigning as an Oceltip director and wanted to appoint Wotherspoon in his place.

Higgins decided to try to get copies of the company register, appoint a secretary and draft a shareholders' agreement, and wished to meet Wotherspoon before agreeing to his appointment. It took some months, but finally, in March 2012, he did get to sit down with Wotherspoon, in Tinkler's Martin Place offices in Sydney, along with Philip Christensen and Paul Flynn, whom he already knew from their Custom Mining days; Tinkler himself wasn't around.

Higgins said he'd been disappointed in Tinkler's email, given all they'd been through together, and he wanted to meet him and clear the air. Afterwards, he emailed Wotherspoon: 'good to finally meet you, and I enjoyed seeing Philip, and later Paul, again. It was just like old times.'

A month later, having heard nothing more, Higgins called Christensen and was shocked to learn that the entire Sydney office of the Tinkler Group had been closed down and its staff sacked, including Wotherspoon, Flynn and Christensen himself. Christensen said he was serving out a notice period but he would be finishing up soon, and that Higgins should call Troy Palmer.

Higgins tried to contact Tinkler but didn't hear back. He would spend the next few months trying to reach his co-director. He contacted Peabody, by now the operator of Middlemount, to work out where the money was going. They were reluctant to get drawn into a dispute between Tinkler and Higgins, and decided to pay all royalties into a trust account until they received a direction signed by both men.

Eventually, Higgins was told that a new account for a company called Oceltip Investments had been set up in June, and that Middlemount had been asked to pay three-quarters of the royalty into it. Higgins was outraged that royalty payments which should have been made to Oceltip were now being paid to an entity he had no links with, and all without his approval.

Higgins was growing more and more concerned. He learned that he was a signatory to another Oceltip bank account that had been opened without his knowledge or approval. When he inquired of the two banks – Westpac and NAB – that had held Oceltip accounts and got copies of the statements, he found out that royalties had been paid into them as early as 2009 (which was odd, as the mine hadn't then opened), and that money was going out to various Tinkler Group entities, including one transfer of $120,000 to 'TGHA' and numerous transfers to the Tinkler Family Trust. It also seemed that Oceltip was late in paying tax, and had built up a debt to the ATO that Higgins could be held liable for.

By now, Higgins was alarmed. He tried again to contact Tinkler. Increasingly stroppy legal letters began going back and forth between Higgins and Palmer. Higgins gave a day's notice that he would be at the Newcastle office listed as Oceltip's principal place of business – the very same Boardwalk building that Higgins and Tinkler had started out in, in the Custom Mining days – where he intended, as a director and shareholder, to photocopy the company's records. Palmer told him he was overseas and would not be around. Higgins went anyway and found that there was no answer to the intercom for Suite C8 (which in fact is listed as the office of many of Tinkler's companies). It was a secure building and he could not get in. He left mail in the external letterbox for Suite C8.

In early November, Higgins started going there regularly to see whether the office was ever attended. He saw nobody and the buzzer was never answered. He tried writing, to convene a formal meeting of directors with Tinkler. Palmer wrote back to Higgins' lawyer that Tinkler would not be available, and added: 'It is no secret that Nathan and Matthew Higgins do not respect each other and have no desire to continue to deal with each other regarding the affairs of the company.'

In December, Higgins went down to the Boardwalk for the tenth time and managed to get inside and up to level three, but found no suite marked C8. It appeared to have been joined with the two adjoining suites, both of which were vacant: C7 was completely empty and had no signage, while C9 had a Patinack Farm sign on the door but no furniture and no people. Another sly Tinkler joke, perhaps. The beating heart of the old Tinklertown was no more.

Palmer had written a month earlier, hoping to set aside concerns over governance of Oceltip and concentrate on selling the Middlemount royalty to Noble, so that Tinkler and Higgins could part ways. But Higgins would not contemplate a sale until the Oceltip accounts and governance were sorted out. At the end of November, he filed in the Queensland Supreme Court against Tinkler, Oceltip Investments and Oceltip itself, seeking repayment of all money that had been diverted from Oceltip without authorisation.

Tinkler's lawyers tried to move the proceedings to New South Wales, stalling the dispute for months. Higgins commissioned a forensic accountant's report, which calculated that some $811,765 of unauthorised transactions – improper diversions by the Tinkler Group – had taken place, and that was *after* factoring in that some of the money taken out of Oceltip's accounts represented legitimate dividends owing to Tinkler (in effect, his share of the royalty up to the point at which he established his new account for Oceltip Investments). The forensic accountant also found that the Oceltip accounts didn't balance, that Oceltip had fallen behind in its tax obligations, that the Tinkler Group had paid money to itself in priority to Higgins' interests, and that the diversion of monies had accounting, tax and cash-flow implications for Oceltip.

Tinkler's lawyers quibbled over the figure – they thought it was $787,943 – but accepted that the money had been removed without authority. They claimed it was a loan from Oceltip to Oceltip Investments. If it had been a loan, Higgins' lawyers immediately asked, what were the terms, what was the security and why hadn't interest been paid at normal commercial rates?

According to Tinkler's calculations, and if it was accepted that Tinkler had the right to set up a new Oceltip Investments account to receive his share of the royalties, the amount 'borrowed' from Oceltip fell to $245,343. Not a huge amount of money, perhaps, but Oceltip was in the early stages of receiving what would likely be more than $100 million in quarterly income over the expected 20-year life of the mine. If they didn't get things properly established from the outset, imagine the disputes later! It was the lack of any proper governance over Oceltip that mattered to Higgins. Worryingly, as Tinkler's press coverage went from bad to worse, Higgins had to consider that there might be no money left to repay Oceltip, which would then be forced to line up with all Tinkler's other creditors.

A mediation was set down for March 2013, but, true to form, Tinkler did not bother showing up. A few days later, however, he suggested coffee in Sydney with Higgins on the morning of 15 March, right before his second day of examinations by the liquidator of Mulsanne. According to Higgins, the meeting was cordial but resolved nothing.

At last, in early May, the dispute was settled. Higgins got back every cent, but it had cost him dearly. Higgins had wanted his money back, but more importantly he wanted to be treated with respect as a co-director and shareholder of Oceltip. Instead, Tinkler had treated his oldest partner with contempt.

For Tinkler, the dispute was more than an unnecessary irritation: it had prevented him from selling the Middlemount royalty at a crucial time, and the consequences of this would prove severe.

*

Through all Tinkler's travails during 2012, nobody knew the true extent of his debt. The *AFR* had reported that it was about $450 million at the end of 2011, and there was speculation that it had risen as high as $500 million. There was a steady drumbeat of speculative reports that Tinkler faced a margin call, which his spokesman Allerton repeatedly and explicitly denied.

The short-sellers were all over Whitehaven at this stage, looking for an opportunity to jump in before Tinkler's stake was sold out from under him and dumped onto the market. While I was working on the *Sydney Morning Herald*'s business desk, I was contacted by a private detective, working for a US hedge fund, who tried to pump me for any intelligence I might have. I tried to work out how much they knew. It was a strange dance, a stalemate: we soon lost interest in each other.

At the end of August 2012, I tried to do my own comprehensive search of the charges registered with ASIC by Tinkler's private companies. After adding up all the loans I could get my hands on, the total figure I got was $638 million. I emailed Tinkler's spokesman Tim Allerton. In half-a-dozen bullet points I explained how I had reached that sum, and asked him to point out any errors. The response Allerton shot back, within an hour, was that Tinkler's true debt was a 'mere fraction' of that sum, but he could not or would not point out where I'd gone wrong.

Tom Reilly and I reported the estimate in the *Herald*, along with Allerton's denial, which would later be proved an outright lie – although, from Tinkler's perspective, it was no doubt a necessary one. With Whitehaven's shares trading as low as they were,

Tinkler could simply not afford to have the true extent of his debts known in the public domain. Everyone would see how perilous his finances were. He was close to underwater.

I stood by my debt estimate, however, and kept reporting it, along with Tinkler's denial, and gradually other media started using the figure. He had hundreds of millions of dollars of debt, and the combined total didn't have a four in front of it, or a five, but a six. Then, suddenly, it was a seven. On 28 November 2012, Bloomberg broke the news that, according to three unidentified sources, Tinkler faced a US$200-million 'bullet' repayment; his total debts stood at as much as $700 million, including accumulated interest. His lenders, including Farallon, could shortly seek to take control of his stake in Whitehaven.

The Bloomberg report sparked a frenzy, sending other business media into overdrive. I called all my best Tinkler contacts, including some fresh calls to a few I'd been chasing for months but had not yet heard from. Amazingly, I got a return call from someone in a position to know the truth about Tinkler's debt.

I was stunned to hear the terms of Tinkler's main loan from Farallon, spelt out precisely: principal, tenor, interest rate. Yes, there had been a US$200-million bullet repayment due at the end of October. Tinkler had already missed it. Breach one. Yes, the loan did include a 'security coverage ratio' – just like the loan-to-valuation ratio in a home loan – and with Whitehaven trading at $3, Tinkler was well under it. Breach two. The whole loan was repayable in a year – at the end of April 2013 – and there was no way known that Tinkler was going to be able to raise the money. Lastly, I was told that Tinkler had just sacked half a dozen of his most senior employees in Brisbane and Sydney, marching them out of the building, ably assisted by his ex-SAS security guards.

After verifying what I'd been told with other trusted sources, there was no doubt that we had a front-page story for the coming

Saturday: Tinkler was clearly in default of his major loan facility. In effect, he was already broke.

Again I emailed Allerton, at 10 a.m. on the Friday morning, with the key financial details I'd been given, along with the sackings. I indicated that we were proposing to publish this in the next day's paper. The response came back within minutes: 'no comment mate'. Then, at four p.m., Fairfax lawyers were given an hour to appear in the Victorian Supreme Court. Tinkler's lawyers were seeking an emergency injunction, on the grounds that I had obtained financial information through a breach of confidence. From the details in my email, Tinkler's lawyers assumed I had been given a copy of the loan. Tinkler himself believed that very few people knew these details. The injunction was granted, pending a fuller hearing in mid-December.

For the next three months I was tied up with Fairfax's lawyers, as we tried to argue for the right to publish, and fought the efforts of Tinkler's team, who were seeking to identify my sources and subpoena my files. Like any journalist worth their salt, I flatly refused, risking contempt charges, and made it clear that I was prepared to go to jail if that was what it took to protect my sources.

It wasn't the usual defamation action. Aside from a minor detail about the conditions under which penalty interest rates might be charged, there was no question of the accuracy of what I'd been told. Tinkler's lawyers themselves were aware that, unlike all the previous reporting, which had been speculative (apart from the Bloomberg story, which came close to the truth), my email was factual and gave a 'very clear picture' of the agreement and its terms.

All the while, Fairfax was restrained by the terms of a super-injunction given by Justice John Digby, which was so broad that, if interpreted literally, it would have prevented every paper the company owned – the *Sydney Morning Herald*, the *Age*, the *AFR*, the *Newcastle Herald* and other titles in print and online – from publishing anything about Nathan Tinkler's financial situation, even material that was already in the public arena. With Tinkler in the news every other day, it was an intolerable situation for a media organisation, and we decided on an unhappy work-around, publishing only those financial details that had already been reported before the case was brought.

It was surprising, to say the least, that we were restrained from publishing confidential information that was clearly in the public interest. The leak, after all, is a tradition in journalism. The law of confidence, however, is a serious barrier to press freedom in Australia – much more serious, in the view of some media lawyers, than the law of defamation – and Minters' legal advice was that Fairfax would lose if we contested it. In the end, a settlement was reached which freed us from the super-injunction but prevented publication of my story.

The lifting of the Tinkler injunction, in March 2013, was reported widely. Fairfax lawyer Peter Bartlett said the super-injunction issued by Justice Digby went further than even Tinkler's barrister had asked for. He described it as a 'dark episode for freedom of speech'.

In his judgement of 21 December 2012, finally made public, Justice Digby had accepted the argument of Tinkler's lawyers, that the facts I'd obtained were commercially sensitive and that their disclosure was a breach of confidentiality. Publishing my story would disadvantage Tinkler by revealing internal financial arrangements, the judge ruled. 'The identified and specific potential detriment to [Aston Resources] and [Boardwalk Resources] and separately also the damage to reputation to [Nathan Tinkler] and the Tinkler Group as a result of the statements ... being published ... outweigh the public interest in freedom of expression.'

My case came amid an apparent business assault on press freedom in Australia, and the journalists' union, the Media Entertainment and Arts Alliance, took up the cudgels: Gina Rinehart was suing Steve Pennells and Adele Ferguson to reveal their sources; powerful Chinese-Australian property developer Helen Liu was likewise suing Richard Baker, Nick McKenzie and Philip Dorling over a series of stories alleging she had bribed former defence minister Joel Fitzgibbon; McKenzie and Baker were being separately sued by the defendants in the 'banknote bribery scandal', which alleged that the Reserve Bank of Australia companies Securency and Note Printing Australia had bribed foreign officials to win contracts. All five journalists faced jail.

Exasperated, I spoke to Melbourne's 774 ABC Radio, telling the veteran broadcaster Jon Faine – a champion of press freedom – that 'I can't tell you what I've learned about Nathan Tinkler's finances'. My story was never published.

But the truth will out, and it wasn't long before the world was aware of the full extent of Tinkler's debts.

*

In December, the Patinack dream shattered. October's reduction sale of 300 broodmares and racehorses had raised a paltry $3 million. Buyers were feasting on Tinkler's misfortune. It was not a great time to be selling broodmares in any case, being the middle of breeding season. Patinack insisted it was not a fire sale, but these were fire sale prices. 'He's in strife all right – he's got to sell,' one happy buyer told the *Daily Telegraph*. One pregnant mare with a good pedigree went for $2000, giving its new owner a free foal.

Worse, Tinkler had not raised anywhere near enough to pay back Gerry Harvey's $20-million loan. Harvey would drive a hard bargain indeed. All Too Hard, Patinack's great hope, would be sold to the Vinery Stud in the Segenhoe Valley – the oldest horse-breeding farm in Australia, which Harvey co-owned – as part-repayment. Under the $28-million deal, Vinery would also take over the former Bowcock property at Segenhoe, which Patinack had bought back in 2008. Rubbing salt into the wound, Vinery also took Onemorenomore, one of Patinack's original Group One winners and now a promising stallion.

The loss of All Too Hard was a critical blow to Patinack. Vinery's general manager, Peter Orton, brimmed with excitement: 'This horse is the real deal. Not only is he a super racehorse, his pedigree is perfectly suited to our conditions. All Too Hard is a spectacular individual physically; he has great power, precocity and presence, all the attributes that make a great stallion.'

Tinkler left Troy Palmer to comment, but he didn't say much. After winning a few more Group One races, All Too Hard went on to stand at Vinery for $66,000.

10.

MULSANNE
BILLIONAIRE IN THE DOCK

The wind-up proceedings brought against Mulsanne in September 2012 had set in train a legal process – linear, plodding, unstoppable – which slowly ratcheted up the pressure on Tinkler, until it became intolerable. Through his spokesman, Tinkler insisted that whatever happened with Mulsanne would have no financial impact on the rest of the group.

True, Mulsanne was just a shelf company, with no other shareholder than Bentley Resources in Singapore (in effect, another shelf company), and no substantial creditor other than Blackwood. But if Tinkler or the company's co-directors – Troy Palmer and Matt Keen – were found to have engaged in insolvent trading, they could be held personally liable for the company's debts, or face a ban from acting as a director of an Australian company. If fraud was established, they could go to jail. Either way, the legal ramifications for Tinkler's business empire were severe, and the damage to what was left of his business reputation could prove irreparable.

Twice the wind-up proceedings brought against Mulsanne were adjourned, with Blackwood's consent, to allow more time for a settlement, but each time, despite a flurry of last-minute activity, the deadline came and went without Tinkler stumping up the $28 million he'd committed to pay. It was understandable that he didn't want to complete the placement: with Blackwood now trading below 15 cents, and with 185 million shares in the market, he could buy the whole company for less than he now owed. But a deal was a deal and he was on the hook.

Amazingly, given everything else going on, Tinkler got close to solving the Blackwood dispute in October 2012. Complaining to Will Randall about the vigorous media campaign against him, which had 'resulted in the termination of five employees of the group in the last week', Tinkler proposed a new deal in which another of his Singapore vehicles, Cayenne Coal, would borrow $100 million from the US investment bank Jefferies and mount a full takeover of both Blackwood and Endocoal. Jefferies would take security over Blackwood and over Tinkler's share of the Middlemount royalty. Tinkler told Randall: 'Higgins is a seller he just wants an offer put in front of him.'

The offer was put to Blackwood on 5 November, a day before the parties were due in court. At first, on behalf of the board, Barry Bolitho declined, describing the proposal as highly conditional and having no financial certainty. Tinkler took exception, warning Bolitho that 'there is a path through this and it is a much better result for shareholders than the one you are headed down'. This was enough to convince Blackwood to agree to seek a week's extension from the court.

Tinkler then wore Blackwood down, to the point that a recommended bid was on the table. He dropped the due diligence requirement and gave a personal guarantee for the funding – in short, he did all he could to get the deal over the line, before the next Mulsanne court date on 14 November.

Tinkler's lawyers were already advising that there were no grounds for him to oppose a wind-up order. As the negotiations dragged on, the lawyers were crossing their fingers that a second adjournment would be granted. Senior deputy registrar Jennifer Hedge was clearly frustrated: 'I'm not keen to adjourn it just every week.' But a final extension was given until 20 November.

As the deadline loomed, there was one thing Tinkler couldn't deliver: an agreement with Higgins over how to dispose of the Middlemount royalty. For Bolitho, Higgins' consent was crucial; he reminded Tinkler that 'only you can procure this'.

Tinkler replied:

> I have no relationship with Higgins no one does and I have been trying for 3 months to do that on the noble deal. It simply won't happen, understand your are trying to be 100% comfortable but it just isn't that situation particularly for me
>
> Maybe get more commercial lawyers next time these guys are a joke. They [Blackwood shareholders] simply do not understand the downside …
>
> I can't do anymore. I have moved and your stupid lawyers take and hold the position. Just not commercial and as I said I can't do more.

By the eve of the court date, however, Tinkler was back at the negotiating table, and terms were almost agreed. Bolitho advised that legally binding documents were due by nine a.m. the following day, and that his lawyers had been standing by for four days.

> TINKLER: I think your lawyers were still looking to press fraud charges against me 4 days ago but I am sure both sides will do everything in their power … My team is ready to go. I don't think its that far away, there is no material change in positions here.
> BOLITHO: There has never been any question of fraud charges. Your team will need to spend many hours on this tonight. It can be done.

At 3.30 a.m. that morning, however, somebody on Tinkler's side who was poring over the fine print took exception to a clause improving Blackwood's security over the Middlemount royalty. This triggered another Tinkler spray:

> [Terms were agreed] until your lawyers again changed the position. That being you would have second ranking security behind noble.
>
> Once again I have done all I can but if you can't control these guys their is little point.
>
> A huge waste of everyone's time. I have reached commercial agreement with you 10 times in the last week and everytime it's fallen over because of your lawyer.
>
> Done all we can our side, huge loss for Blackwood shareholders but it's not on us. Your lawyers went home, not mine. Noble is your major shareholder and you have security [behind] them. I cant see how you are not aligned. Shareholders have a security blanket under this scenario, they have nothing tomorrow; can't believe your lawyers can't tell you that.

Blackwood had no choice. Mulsanne had now been given five ultimatums over a four-month period. Will Randall's prediction that full due process would have to be exhausted was coming true. On 20 November 2012 the Supreme Court finally ordered that Mulsanne be wound up, appointing Ferrier Hodgson as liquidator. Tinkler was on the ropes.

From the outset, Ferrier's liquidator, Robyn Duggan, said the investigation would take months. At first blush, there didn't seem much to investigate: Mulsanne had an undisputed debt and no assets to speak of; it appeared an open-and-shut case of insolvent trading. But the investigation that followed would lift the lid on Tinkler's foundering business empire.

*

In December 2012, as the rest of Tinkler's empire seemed to be collapsing, Duggan's first 'report as to affairs' confirmed that Mulsanne was 'under-capitalised' (liquidator-speak for broke). On 7 January 2013 Duggan got orders to produce all relevant documents against the three directors, including their personal financial affairs, and against the company secretary, Aimee Hyde, as well as summons for them all to appear for liquidator's examinations in March. Tinkler had blamed adverse market conditions for the failure of the Blackwood placement. Duggan's report to creditors in February flagged possible insolvent trading and unreasonable director-related transactions that required further investigation.

Providing personal and company files to Mulsanne's liquidator was bad enough. But for Tinkler, being forced to appear on the witness stand – where he'd be cross-examined under oath about his financial affairs, in full view of the nation's media – was a nightmare come true. He would do anything to avoid it.

The true extent of Tinkler's last-ditch efforts to avoid taking the stand would not be revealed until the liquidator's affidavit was tendered in court in March 2013. Once again, Tinkler left it until the last minute, and by now his bargaining position wasn't great. At the end of December he'd missed the due date on a $5-million, six-month loan that Noble had discreetly given him at the end of June 2012, secured against his Middlemount royalty. Now they were in dispute and, with Tinkler on the ropes, Noble had declined to extend the loan.

Somehow Tinkler found the money to pay out Noble in January 2013, and was soon again proposing a takeover of Blackwood by Cayenne Coal. In line with the fallen coal markets, and with Blackwood's shares now trading below 12 cents apiece, the offer price was lower this time: just $15 million. Cayenne would issue a notice of an intention to bid, which would be binding under the 'truth in takeover' provisions of the corporations law, and Blackwood would stay the wind-up proceedings and cancel the liquidator's examinations.

Tinkler threw every unencumbered asset he had at the deal, including the Middlemount royalty and even his dad's stud farm, Serene Lodge, which would be security for $6 million. Once again, Blackwood's Barry Bolitho was contemplating a last-minute takeover proposition, and once again he indicated it would be supported if there was evidence that Tinkler could pay. Was there an independent valuation of Serene Lodge? Was there authorisation from Higgins that the royalty could be used as security?

By now, trust between the parties was so shattered that Blackwood's lawyers wanted screenshots from Tinkler's lawyers' bank account, showing the bid funds sitting in escrow, before it would agree to any further extensions. Desperate emails flew back and forth until 10.30 p.m. the night before the examinations were due to start. They were still flying at 10 a.m. the next morning, when Blackwood consented to putting the hearings back until two p.m. The closed-door talks continued.

The money still wasn't there, however, and the examination hearings reconvened that afternoon. Unbeknown to all but his lawyers, Tinkler wasn't there either. When he was called to appear, his barrister Alec Leopold, switched tactics, desperately trying to stay the proceedings on the basis of an entirely new argument: it was an 'abuse of process' for the same lawyers to act for the liquidator Ferriers, the major creditor Blackwood and the major shareholder Noble.

The liquidator's barrister, the heavy hitter Robert Newlinds SC, branded the stay application 'half-baked' and a 'forensic stunt, the type of which I must say Mr Tinkler, at least in the press, has a reputation for'. Newlinds told the registrar: 'We are not at all convinced that Mr Tinkler has ever had any intention of complying with the summons and there's an easy way to test that: is he here?'

The air thickened: it would be a grave admission if Tinkler had defied the court's summons. The absent billionaire was seconds away from having an arrest warrant issued against him. Five times Leopold refused to answer Newlinds' question, until the registrar himself asked whether Tinkler was in the court precinct. 'No,' answered the embarrassed Leopold.

> NEWLINDS: What about in the City of Sydney?
> LEOPOLD: We say we shouldn't be interrogated about that in the circumstances of this application.

Newlinds warned Leopold: 'We want to make it 100 per cent clear that if Mr Tinkler … is not here next Thursday we will be applying to have him arrested.'

That afternoon, Tim Allerton rang round all the key journos covering Tinkler to stress that, while his client was not in court on Friday, he was in a hotel room two minutes away and available to attend if ordered. Two days later, however, when the ludicrous 'abuse of process' application was heard – and was duly dismissed – Newlinds called Tinkler's solicitor, Scott Harris, to the stand. Under oath, Harris had to admit that his client had been in Singapore the previous Friday. Outside court, Allerton – just doing his job, of course – explained that he was now told that Tinkler had been in a hotel near Singapore's airport, ready to fly to Sydney if needed!

On the Thursday morning, 14 March 2013, the assembled media were outside the Supreme Court early, to try to catch Tinkler on his way in. It still seemed unlikely he would turn up. It is rare to get a billionaire – or a former billionaire – in the dock. The two packs, stationed either side of the building, assumed he would arrive by limo.

He surprised us all by walking straight through Hyde Park with a bag over his shoulder, and went into his barrister's chambers next door. The photographers ran over and got plenty of shots of Tinkler, a lot trimmer than in his last public outing, and with his head shaved. He had a beaming smile for the cameras.

*

The courtroom, of course, was packed to bursting. For the next day and a half, Tinkler played nice as he was grilled by Robert Newlinds. Softly spoken, and asserting legal 'privilege' before every answer to avoid incriminating himself in any later prosecution, Tinkler could not have been more helpful – except that his whole argument was based on one-sided recollections of undocumented conversations.

To avoid a charge of insolvent trading, Tinkler needed to show that he had reasonable grounds as a director to believe that Mulsanne could meet its liabilities, as and when they fell due – particularly in July 2013, when Blackwood expected its $28 million. Tinkler argued that he was going to get most, if not all, of the money from Noble, which had agreed to buy his Middlemount royalty ... sort of. The deal was never put in writing, he said, but was based on a series of conversations he'd had with Will Randall, starting round Christmas 2011.

What Tinkler was saying was plausible. Noble was in the process of 'recycling' its Australian coal assets, switching to an 'asset light' strategy as commodities markets got choppier – the 'new normal', according to its founder, Richard Elman – and securing some predictable cash flow from a royalty was very attractive. Tinkler had benefited handsomely from Noble's desire to get its foot on coal assets in 2007, when prices were about to hit all-time highs. Five years later, the tide had turned for Noble.

Noble knew the Middlemount mine intimately and was a logical buyer of Tinkler's royalty – in fact, it had always been keen to buy it. Noble wasn't exiting Australia completely: it had invested a small amount of money to take half of Blackwood, and Randall was talking to Tinkler about his grand plans for the exploration company.

The problem with Tinkler's argument was that he had no evidence to back it up. Plenty of correspondence from the time had been provided to the liquidator, and up until October 2012 none of it had so much as mentioned the Middlemount royalty. Tinkler, on oath, was insistent, though, and spoke in some detail about how he generally kept diary notes in a pad – he'd run through one or two a week – which would then be filed by date. But it seemed there was no record of his many discussions with Randall.

This lack of documentation frustrated Newlinds no end:

NEWLINDS: Can you explain why it is that there are apparently no other notes ... at all recording any discussions between yourself and Mr Randall or anyone else about the sale of the royalty stream?
TINKLER: Privilege. I'm unable to at the moment ...
NEWLINDS: Do you recall making any such notes at the time?
TINKLER: Privilege. I do think, I do think there are these notes – there are some notes that exist, I do think so.
NEWLINDS: Do you recall giving any of those notes to your solicitors for the purpose of answering the orders for production?
TINKLER: Privilege. I recall looking for them ...
NEWLINDS: ... but did you find them?
TINKLER: No, I wasn't able to find them ...
NEWLINDS: And you can't find anyone that's got any note of this ongoing dialogue between yourself and Mr Randall and anyone else at Noble, can you?
TINKLER: Privilege. I haven't been able to find those direct notes.
NEWLINDS: On what you're telling us there were an awful lot of those discussions.

TINKLER: Privilege. There were a lot of discussions.

NEWLINDS: And so you know that regardless of whether the orders capture it or not, it would assist your position if you produced those documents to the liquidator?

TINKLER: Privilege. Yes.

NEWLINDS: And may we take it that you've looked hard and long for them?

TINKLER: Privilege. I would say that I personally – I wouldn't say that I personally looked hard and long. I would say that my team has under instruction and, you know, that perhaps there's something that has been missed.

NEWLINDS: Those searches no doubt will continue?

TINKLER: Privilege. They're ongoing, yes, as we speak.

NEWLINDS: Because you understand don't you that if it be right that your ordinary business practice was to make notes of important matters that you were dealing with from time to time and to keep those notes, then the question of the sale of the royalty stream and where the $28 million was coming from inevitably had to find its way into those notes, didn't it?

TINKLER: Privilege. I would suggest to you that, you know, the better I know someone and the more I know them, the less likely I am to be taking – writing a page full of notes in a 15 minute conversation, so obviously as I'm learning about things and building relationships I make a lot of notes. When I'm working with somebody that I've known for five years and call freely and speak to several times a week, that's very different.

Tinkler was painting a picture of the open working relationship he had with Noble, and with Randall in particular – so close that a $28-million deal between them could go entirely undocumented, right up to the last minute and beyond. Tinkler operated in the business stratosphere: a rarefied world in which detail, execution and even the law were fussed over by minions, while verbal agreements were good enough, between trusted parties, to move vast sums of money around at will.

As the deadline for settling the Blackwood placement came and went, Tinkler said, he had no doubt that a deal could still have been done with Noble – at a moment's notice, if need be – and no doubt Randall could make that happen, if he wanted to.

> NEWLINDS: Did you think he had authority on his own without going back to the board of Noble to enter into such an agreement?
> TINKLER: Privilege ... I have done transactions with Will Randall in hundreds of millions of dollars, I would have thought that one at twenty to thirty million dollars was not going to be out of his realms of authority.

In fact, it is clear from his testimony that the balance of power between Tinkler, the so-called coal baron, and Randall, the coal chief at Asia's largest commodities trader, had shifted uncomfortably. While Tinkler was struggling to find $28 million for the Blackwood placement, Randall was an executive director on the main board of a company that had 15,000 employees and turnover exceeding $90 billion.

> TINKLER: Mr Randall just went off-line.
> NEWLINDS: So you were trying to see if Noble would lend you the money so that you could buy shares in the company that Noble was already a shareholder in ... and the person you were trying to speak to on that topic at Noble, wasn't even speaking to you?
> TINKLER: Privilege. Would not answer his phone.
> NEWLINDS: May we take it that you pretty quickly worked out that he wasn't interested?

TINKLER: Privilege. I quickly worked out that I'd been hung out to dry, yes.

All the conversations with Randall had come to nought. It was Tinkler left dangling, forced to recognise that Randall had bigger things on his plate than him or Blackwood. It was Tinkler trying to preserve a relationship with Randall.

Newlinds asked Tinkler why, in his emails to Blackwood around the settlement date, he did not point the finger at Randall:

NEWLINDS: Can you explain why in this email you didn't say to Barry Bolitho, Will Randall and everyone else listen can you call off the dogs because you well know that I've been relying on a sale of the royalty stream to Noble to get the money to complete this transaction and I've had numerous discussions with Will Randall starting as long ago as Christmas wherein he's assured me that that would all take place and he's let me down and I'm hanging out to dry. Can you explain why you didn't say something along those lines?
TINKLER: Privilege. Because I was trying to protect mine and Will's working relationship … the last thing I wanted to do was paint a terrible picture of Noble letting me down to the Blackwood directors as major shareholders and what goes on you know me and my relationship with Noble as being a failure … [we'd had] our bumpy bits before in commercial negotiations and so forth and this was just another one.

Then Newlinds asked why, in October 2012, when Tinkler finally did offer the Middlemount royalty to Noble in writing, he made no reference at all to the earlier discussions he'd supposedly had with Randall, going all the way back to Christmas.

TINKLER: Privilege. In my experience, at this level of business, the best way to ask for help is not by kicking someone in the teeth, so I have a personal relationship with Will. My business relationship with Will and Noble will continue beyond this and we'll go – I didn't see a need for documenting emotion and the issues that I have in a formal letter like this that Will had asked me to put forward in support of Noble acquiring the royalty … I was still hopeful that would happen. I wanted to move forward and build value in Blackwood and put this behind us. I very much wanted to work with him as you can see from the tone of the letter, so I didn't want to – I didn't see the need to berate him.

Randall, of course, was never examined by the liquidator, so we will never get his side of the story. Those who know Tinkler say the episode is typical, and provides an insight into why he repeatedly buys the next asset before he has the money. Tinkler was not lying about his conversations with Randall, but he was clearly overconfident that a series of informal discussions represented an agreement good enough to take to the bank.

It was a classic Tinkler overreach. In his mind, the deal with Randall was done. Whether it was 100 per cent or 75 per cent of the royalty was for Randall and Higgins to sort out. Mentally, Tinkler was already spending the proceeds of the sale of his share.

He was not naive about Noble, though, and as settlement loomed and the Blackwood placement turned into a headache, he realised he was going to have to try to raise the money elsewhere.

NEWLINDS: Well, what about this morning when you were telling me about all these financiers that you were talking to in connection with raising the money? You had other options, didn't you?

TINKLER: Privilege. When you work with Noble you always have other options.
NEWLINDS: Is that because they can't be trusted, is that it?
TINKLER: Privilege. That's – how do you think I got here?

By buying it low when the markets were off, and by selling it high when they recovered, Blackwood was to be Tinkler's next big coal play: roll in Endocoal, pick up Integra off Vale, package it all up and sell it to some Indian company for a billion dollars. 'The whole reason to do the share placement agreement was to grow a company,' Tinkler told the courtroom. 'We weren't doing it to stand still.'

But the bounce never came – or not in time for Tinkler. As he told Newlinds, by the middle of 2012, sentiment in the coal market had 'gone off a cliff'. The dream was over. There were to be no more freewheeling mining deals on the world stage. Tinkler was a coal baron no more. Even at this point, though, while sitting in the witness box, he was in denial.

*

On the second day of Tinkler's evidence, Newlinds grilled him about his personal finances, to determine whether he could pay any judgement against him. Everyone's ears pricked up. It was titillating to hear that Tinkler had declared taxable income of just $9834 in 2010–11, mainly interest from various bank accounts. Once again, a billionaire was revealed to be paying no tax at all.

Apart from a farm in Port Macquarie worth about $700,000, no assets were in Tinkler's name – no other real estate, no cars, no stocks or bonds. Tinkler had a couple of joint bank accounts with his wife that had roughly a quarter of a million dollars in them. That was it.

It emerged that Tinkler got all his spending money by way of tax-free distributions from a unit trust controlled by his wife,

Rebecca, as the unit holder *and* trustee. He identified it as the Tinkler Group Family Trust. Created in 2007, it had sold his shares in Macarthur Coal, and had paid around $100 million in tax.

Company records show no such thing as the 'Tinkler Group Family Trust'. There is a Tinkler Group Holdings Family Trust, and there is a Tinkler Family Trust (later renamed the Oceltip Family Trust). If he'd forgotten the name, Tinkler also had no idea how much money he'd taken out of the trust in 2010–11.

> NEWLINDS: If we rounded it to the nearest 100,000 would that help?
> TINKLER: Privilege. No.
> NEWLINDS: If we rounded it to the nearest million would it help?
> TINKLER: Privilege. I couldn't be sure.

In fact, Tinkler seemed not to know how any of it worked:

> NEWLINDS: How do you get drawings out of the trust?
> TINKLER: Privilege. I'm not sure, I'm not exactly across those, across those issues.
> NEWLINDS: So is this what happens, your wife gives you money from time to time?
> TINKLER: Privilege. I'm very lucky, yes.

This got a laugh: Tinkler was charming the room, playing dumb. It was understandable that he might be foggy about the details, but when he was asked to describe the assets and liabilities of his wife's trust, Tinkler departed from credibility.

Until this point, everyone had assumed that Tinkler's most valuable asset was his 19 per cent stake in Whitehaven. For months journalists had been writing about the falling value of that stake –

at $3 per share, it was worth about $600 million – and comparing it to the value of his reported debts against that stake, an estimated $700 million, with the obvious conclusion being that Tinkler was underwater on his major loan, and therefore close to broke. Newlinds tried to put his finger on this problem:

> NEWLINDS: I don't want to over-simplify things, but does the wealth of the Tinkler Group of Companies depend on the share price for Whitehaven shares at any particular time?
> TINKLER: Privilege. Not really.

If true, that was a bombshell. What else did Tinkler have? On oath, Tinkler said the total amount of assets in his wife's trust was 'around $1.4 billion' – which made headlines the next day. (In a later exchange, Newlinds erred, calling it $1.2 billion; Tinkler seemed comfortable with either figure.) When asked what was in the trust, apart from the Whitehaven stake, Tinkler identified Aston Metals, Hunter Ports and Hunter Rail, and the Patinack Farm stud, which he reckoned was worth 'at least' $100 million after debt.

To anybody who knew these assets, the valuation was laughable. Aston Metals, bought for $2 million, was now worth $50 million, tops, based on its drilling results and industry estimates. Hunter Ports and Hunter Rail were just ideas, failed infrastructure projects, all but worthless. Patinack Farm was heavily encumbered and had twice been put on the market. With Whitehaven's shares in the toilet, there was no way the collection of assets Tinkler identified was worth anything like $1.4 billion.

Asked to value any of the assets himself, Tinkler demurred: 'Privilege. I guess – I deal in the strategic market, not in the marketplace per se, so I think those valuations are going to be different, but they all have very good value.'

With something like a hundred active and inactive companies in Australia alone, not to mention a dozen trusts which do not file accounts, and the possibility that assets could be squirrelled away in any number of offshore jurisdictions, it is impossible for a journalist to be definitive about Tinkler's true wealth. But if there were other assets than those he identified on the witness stand, Tinkler had not been able to put his hand on them to keep himself out of court.

Valuing illiquid assets such as real estate and infrastructure – or even big blocks of shares in listed companies – can be arbitrary. The value of a debt is much less so. While Tinkler took an extremely generous approach in the witness box to the value of his assets, he vastly played down the extent of his debts, estimating the trust's total liabilities at 'around $600 million'. As would be confirmed some months later, this figure was a serious understatement: his loans from Farallon alone were worth more than that, and there were also loans from Jefferies, Westpac and others.

In fact, while on the stand Tinkler appeared to have lost track of his borrowings, and seemed not to remember that his share of the Middlemount royalty was encumbered: Noble had lent him $5 million against it back in June.

> NEWLINDS: So you would suggest that the trust, controlled by your wife, has a net wealth of about $600 million?
> TINKLER: Privilege. Yes.
> NEWLINDS: And if your wife chose to she could deploy some of those assets to your benefit?
> TINKLER: Privilege. Some of those assets could be used, yes.
> NEWLINDS: In recent times, recent days and weeks, did you ask your wife if perhaps she could release some funds from the trust so that you could deal with some pressing demands by creditors?
> TINKLER: Privilege. No, I have not.

NEWLINDS: Did your wife ask you whether, having regard to what was happening in your life, it would be advantageous to you if she gave you some money?

TINKLER: Privilege. I think it was more around liquidity than actual asset value.

NEWLINDS: All right, I think I understand what you're saying. Is the position that the trust, whatever the valuations of these various assets are, is asset rich but cash poor at the moment?

TINKLER: Privilege. At the moment, so they are not yet cash flow assets.

NEWLINDS: Why didn't you just go to a normal bank and ask for a loan?

TINKLER: Privilege. Normal banks do not understand my business model.

NEWLINDS: So that was out of the question, to go to a normal bank?

TINKLER: Privilege. It's just out of the question for me to deal with normal banks, yes.

NEWLINDS: Why is that?

TINKLER: Privilege. Because I, I – they don't understand how I create my wealth.

NEWLINDS: Well how do you create your wealth? Let's see if we can understand it?

TINKLER: Privilege. I identify, I identify mining assets which I think are undervalued and bring them, you know, prove them up, build value in the asset and then identify strategic partners that would be attracted to those assets.

Nobody understood it anymore. As Tinkler strode from court, trailed by TV cameras and a pack of reporters, the ABC reporter Conor Duffy managed to get close enough to ask a few less polite, more pointed questions:

DUFFY: How much financial pressure are you under Mr Tinkler?
TINKLER: No comment, thanks.
DUFFY: Why don't you pay your bills?
TINKLER: No thanks.
DUFFY: Do you worry you could be the next Alan Bond?

At that, Tinkler just laughed.

11.

DEMOLITION MAN
THE BRIEFEST BILLIONAIRE

Nathan Tinkler might have been buoyed by his appearance at the Mulsanne liquidator's examinations. Over two days he had squared off against one of the sharpest silks in the business, and given as good as he got. After fighting for months to avoid a court appearance, he'd called their bluff: his financial situation wasn't so bad. According to the next day's papers, which ran front-page stories based on the evidence he'd given, he was still a billionaire; at least, his wife was.

Tinkler spent the next weekend at the Melbourne Grand Prix, where his Porsche, driven by Steven Johnson, placed fifth in the second round of the Carrera Cup. Remarkably, Tinkler managed to evade the heavy media contingent at the race completely. He was seen, however, in the close company of a personal assistant. Partly as a result of the testimony he'd just given, Tinkler's private life was now fringing on becoming public. If all his assets were in Rebecca's name, and if he therefore depended on her for his income, a divorce would have many financial implications.

Three weeks after Tinkler's appearance in the witness box, Patinack Farm was put on the market. The whole operation – two studs, eight stallions, 1000 horses and a couple of hundred employees – would be sold through Gerry Harvey's Magic Millions. Nathan Tinkler's childhood dream, to race the progeny of his own stallions, was over. If he still thought Patinack was worth over $100 million, 'at least', as he'd told the court, here was the acid test. Agent Vin Cox told the media that Patinack would sell for more than $150 million on a 'walk-in, walk-out' basis.

At Doomben Racecourse in Brisbane, where his promising two-year-old Hooked had won the opening event, Tinkler told journalists that the racing industry had thrown him 'on the scrapheap'. He had received a couple of serious bids on Patinack, which would sell within weeks, he said, sealing his exit from the industry.

> You can have a lot of fun in this industry but you can't do that without the support of the industry. It's too big a topic to go into at the moment, but let's just say there's bad luck and there's Patinack luck. I didn't see it going like this, but it's the right thing to do. We've got a great team and a lovely bunch of horses. Someone else will have a lot of fun with it. I think I need a break. You never say never. I love the sport dearly, I'm sure I will be back one day.

In quick succession, Tinkler's mansions came up for sale. An auction date was set for the Pullenvale home in Brisbane, while the Ocean Street properties at Merewether were also said to be on the market – although they were now looking much the worse for wear. Even the prized 'Noorinya', at Sapphire Beach, was reportedly for sale, although local real-estate agents denied all knowledge of the listing.

The Demolition Man's house in Brisbane quietly fetched $2.4 million. The receivers were still battling to sell Tinkler's jet and chopper – the sticker price was falling, and talk that he might refinance and buy them back was revealed to be just that: talk. Soon Tinkler's private hangar at Brisbane airport was up for sale as well.

The great Tinkler wind-up, it seemed, was on. As was known only to a few, but suspected by many, some US$634 million of debt to Farallon was falling due at the end of April 2013. In reality, it didn't matter anymore what Tinkler thought his assets were worth; what mattered was whether he could pay off his debts. Whitehaven's shares had been trading ever lower for a year and were stuck below $2.30. At market prices, Tinkler's 19 per cent stake in the company was now worth less than $435 million. There was no way known he could meet his obligations to Farallon.

Tinkler's sell-off spooked Blackwood, as Mulsanne's largest creditor, which was still hoping to recover the $28 million Tinkler owed it. At the start of May, Mulsanne's liquidator had launched claims of insolvent trading and breach of director's duties against Tinkler and his co-directors, who could be held personally liable for the debt. Next the liquidator rushed to court, seeking an injunction to freeze the assets not just of Tinkler but also of Rebecca and of Oceltip Investments, as trustees of the Tinkler Family Trust, until the claims were resolved.

Singapore had turned treacherous: Tinkler was in a vice, crunched by his two biggest backers. Noble, on its way to becoming a commodities giant, had made Tinkler's first fortune, on the Middlemount acquisition. Tinkler's fellow Australian Will Randall had been there for the ride with him since 2007. Now, as Blackwood's biggest shareholder, in Tinkler's hour of direst need, Noble was going for his wife's assets. The multibillion-dollar hedge fund Farallon, and particularly Ray Zage, had made possible Tinkler's acquisition of Maules Creek. Now, Zage held security over almost everything Tinkler owned and was poised to pull the plug on his empire.

Tinkler's fortunes had risen thanks to these two men and their powerful corporate interests. It had suited Randall to boost a junior Australian coal miner like Tinkler in 2007, when commodity prices were rising and when Noble needed coal supplies on terms that were not dictated by the major miners, BHP and Rio. It had suited Farallon to make a mini-mogul out of Tinkler in 2010, when a GFC-induced restructure by the floundering Rio Tinto threw up just the kind of 'special situation' the hedge fund revelled in.

With Noble's money and then with Farallon's, Tinkler had appeared as a bull-market genius: anyone who bought anything looked smart in a rising market. By 2013, with coal prices going nowhere, it no longer suited either Noble or Farallon to be exposed to Tinkler. Both had done well out of him, and both had now had enough.

Tinkler dropped out of *BRW*'s Top 200 Rich List altogether in May 2013, with the magazine calculating that his net wealth had fallen below its cut-off mark of $235 million. It was a stunning fall from grace. In what counts for official figures in Australia, Tinkler had debuted with $426 million in 2008 at the age of 32, and peaked at $1.1 billion at the age of 35 (although, in retrospect, there is no way this was an accurate estimate of his net wealth in September 2011). Two years later, his fortune was gone. By the time the Young Rich List calculations came around again in September 2013, *BRW* reckoned Tinkler's net wealth to be below its $18-million cut-off. In the 30-year history of the list, no one had ever risen so far so fast, and then, equally surprising, fallen just as quickly. He was our youngest billionaire, and our briefest.

But at just 37 Tinkler could not be written off, as *BRW*'s editor, James Thomson, cautioned in a later opinion piece. 'There is no shortage of obituaries for Nathan Tinkler,' Thomson wrote, adding: 'not so fast.' Tinkler was young, he'd built wealth quickly before, he still had assets, he had an eye for undervalued assets and, finally, miners loved a comeback:

> The mining sector has special place in its heart for stories of redemption. Before Andrew 'Twiggy' Forrest had Fortescue,

he had a disaster with Anaconda Nickel. The past of miner and former billionaire Frank Timis features two convictions for heroin possession. Even Alan Bond made a brief return to the Rich 200 courtesy of his mining interests.

For about a month, around June 2013, Tinkler entered a twilight zone. Everything was for sale but nothing had sold. The Pullenvale auction was cancelled, and it was all quiet at Merewether. Patinack was still on the market, but if any genuine bids came, they were well short of the mark.

Even as buyers were supposedly doing their inspections, a massive Patinack reduction sale got underway: some 400 yearlings, broodmares and weanlings would be sold at the Magic Millions, with no reserve prices set. All over the country, buyers started poring over the auction catalogue, checking form and bloodlines, organising veterinarian and dental check-ups, doing their figures. At the sale itself, excitement was at fever pitch. Rapid-fire, the auctioneer hammered the message home: everything must go. Tinkler was nowhere to be seen, but Gerry Harvey strolled about purposefully, from table to stable, even giving the ABC a few thoughtful observations on Tinkler's contributions to the racing industry:

> He's put a huge amount of money in and I don't know how much he's going to end up losing – $100 million, $200 million or some huge amount of money. That money went into this industry and this industry has been a huge beneficiary of that money. I think he's done a lot of silly things but he tried – he tried.

The prices weren't all that bad; Nechita even went for $1.5 million. On the third day, however, it went round the room that Patinack was bidding on its own horses. Patinack withdrew 22 horses from the sale, infuriating potential bidders, many of whom had travelled distances to be there. There was uproar and Magic Millions was forced onto the back foot.

On 26 June the Farallon-led loan syndicate took control of Tinkler's Whitehaven shares. Dressed up as a 'sale' at an agreed price of $2.96 a share, the transaction was more like a lender foreclosing on a borrower in default.

Tinkler's largest asset had been wrenched off him. Chairman Vaile admitted having a 'great deal of empathy' for Tinkler. 'I know Nathan well,' he said. 'I'm sure you will see more of Nathan investing in the coal industry in the future.' Tinkler thanked his lenders and wished Whitehaven well, but his pain was evident from the statement issued on his behalf:

> While we are happy with the price we have received, being a 40 per cent premium to the current market price and a recent high for the company, we feel strongly that this still significantly undervalues the company's underlying asset base. Many will be aware of the emotional attachment that Mr Tinkler has to the assets of the company, specifically Maules Creek, and that selling this stake was a difficult decision. However we believe that no longer being a substantial shareholder of Whitehaven will benefit all existing shareholders.

Farallon and its clients had effectively converted Tinkler's debts to equity, but at the agreed price they would only recover a maximum of $565 million – not enough to repay the principal, let alone the accumulated interest, which, after a year at 15 per cent, would have reached almost US$730 million.

Under the deal, if Whitehaven's shares went above $2.96 within the next nine months, Farallon would count that increase against the balance of Tinkler's outstanding loan. They need not have bothered. Tinkler's exit barely moved the stock, and by late 2013, after the release of the company's full-year accounts, Whitehaven was back below $2 a share. In round figures, Whitehaven shares would need to reach $3.70 by the end of March 2014 in order to wipe out Tinkler's outstanding debts. Possible, but highly unlikely.

Whitehaven's credibility is damaged. The Tinkler saga, the Boardwalk impairment, the tardy disclosure of the problems at Narrabri, the endless delays at Maules Creek – all have taken a toll. Although he is gone from the share register, Tinkler's spectre hangs over the board. One way or another, the chairman Vaile, managing director Flynn, and directors Christensen and Zage all owe their positions to their relationship with Tinkler, and the market remains wary. Even positive announcements – such as Tinkler's exit, or the record tonnages mined from Narrabri – have failed to lift the company's shares. It is too easy to blame the soft coal market. Billions of dollars of value have been wiped out since Whitehaven's merger with Aston. Some shareholders mutter darkly about class actions – you can't smoke that much money and get away with it, they say – but give no indication what the claim might be, or against whom.

Whitehaven's fate remains tied to Tinkler's in another way: Tinkler, Christensen and co. hold a large number of milestone shares, courtesy of the Boardwalk transaction, which will vest if there is any takeover bid for Whitehaven. With the company's shares trading below $2, that is a very real possibility if even the slightest recovery in coal markets occurs. Even at such low levels, the vesting of those milestone shares would be worth some $60 million to Tinkler, giving him a real chance of wiping his slate clean. Tinkler is said to be trying to drum up interest in a new bid for Whitehaven – in China and in the United States – but is struggling to get meetings.

One positive that came from Tinkler's sale of his Whitehaven stake to Farallon was that he managed to free up enough cash to get

Blackwood to drop its proceedings against Mulsanne. Although he owed $28 million, Tinkler agreed to pay $12 million and to relinquish his stake in Blackwood, ending the saga once and for all.

Farallon also held security over all the most valuable parts of Tinkler's business. Aston Metals, with its remote zinc/copper projects near Mount Isa, was the next to be put on the block. Bids were due by October 2013. Its sale should reduce Tinkler's debt somewhat.

His debts to Farallon and others were not the only thing hanging over Tinkler's head. For some time, analysts had speculated that any 'sale' of his Whitehaven stake would crystallise another large capital-gains tax bill – theoretically as high as $160 million – given that Tinkler's entry price into both Aston and Boardwalk had been extremely low. Tinkler consistently denied it but provided no justification as to why he would not be liable for gains tax on these assets.

The amount could perhaps be offset by accrued losses elsewhere in his empire – if he successfully sold Patinack, for example. All his assets were held in a web of trusts. Who knows what tax miracles his accountants might come up with? At the very least, though, and not for the first time, Tinkler would have to do some explaining to the taxman.

*

In 2013 the Newcastle Knights made the finals. This was not a big deal in itself; they had done it in 2011. Back then, though, it had been despite Tinkler, as some commentators said at the time, not because of him, given the turmoil that had surrounded the appointment of Wayne Bennett and the way players were purged to make way for his hand-picked recruits. In 2012 Bennett's 'Dad's Army' of older players such as Kurt Gidley, Willie Mason and Timana Tahu had suffered the humiliation of missing the finals altogether. Compounded with the endless uncertainty over Tinkler's financial situation – the missed pays, the unpaid bills, the wind-ups – and his poisonous relationship with the local media, the Hunter Sports Group's takeover of the Knights was starting to look like a disaster.

In January 2013, though, the club members had come out and backed Tinkler, saying that he had done everything he'd said he would do, that the club was most grateful, and that his bank guarantee could drop back from $20 million to $10 million, as originally planned. Wayne Bennett said he still had confidence in Tinkler; indeed, if Bennett had said anything else, it would have been all over for Tinkler. The Knights were getting on with the job. The vociferous public debate about a 'Plan B' for the Knights, in the event that Tinkler fell over, was replaced by background chatter about how the club might be gently transferred out of the Hunter Sports Group's control, should the need arise. The team grew in confidence in the second half of the 2013 season and made the top eight.

In early September the club announced a new headquarters: the 'Home of the Knights' would now be at Wests' Mayfield club. Wests runs one of the most profitable club operations in the country; although it had backed the rebel Hunter Mariners during the Super-League war of the 1990s, it had also been a generous supporter of the Knights over the years, donating $1 million to the club.

At the press conference to announce the new home, Wests' chief, Phil Gardner, spoke about 're-engaging the Knights with the rest of the community'. But where were the owners of the club, Tinkler and the Hunter Sports Group? Asked about Tinkler's absence, the Knights advisory board chairman Paul Harragon said that while Tinkler was 'a great leader, he's our backbone … he's taken away all that debt on behalf of the community and put the Knights on solid, solid ground', the day was about thanking Wests for its benevolence and vision.

Ten days later, in the first week of the finals, the Knights beat the perennially tough Canterbury Bulldogs, 22–6, knocking the previous year's minor premiers out of the competition. This was something we hadn't seen for a while. When the Knights edged out the 2012 premier Melbourne Storm the next weekend, winning 18–16 after an epic struggle, Newcastle went mad. The Knights were on fire, and Wayne Bennett was delivering. The team was now just one win away from the Grand Final. For the fans, it wasn't a question of whether Newcastle would win its first premiership since 2001, it was how many Grand Finals the club would go on to win in the coming years.

With the excitement building ahead of the Semi Final against the Sydney Roosters, to be played at the old Sydney Football Stadium, the Knights' owner, Nathan Tinkler, turned up to congratulate the team ahead of the big showdown. In decent shape and wearing a white polo shirt, Tinkler didn't do any media but was pictured chatting to the players, especially Mason. 'Good to see,' said Paul Harragon. Amid the finals hype, Wayne Bennett told reporters he was 'elated for Nathan, because without Nathan, this team wouldn't be here today, simple as that'.

But it wasn't to be: the Roosters were too good. Still, Bennett and Tinkler had given the fans something to cheer, and they were grateful.

Would Bennett, now aged 63, stay on? The North Queensland Cowboys wanted him, and there was always speculation that he'd one day return to the Brisbane Broncos. A few weeks later, Bennett, overwhelmed by the support that Newcastle had given him, confirmed he would see out his four-year term. The Knights were back. And while Tinkler's money continues somehow to flow to the club, week after week, his stewardship is also secure.

*

It is often observed that Tinkler's rise and fall tracked the coal price, and that is true. In the simplest terms, his story is bookended by two major flips in world coal markets: he rose when China switched to become a coal importer, and he fell when the United States switched to become a coal exporter. It all happened so quickly. Tinkler was quick to spot the boom that China unleashed upon Australia. But, like the rest of the coal industry, he was slow to appreciate the implications of the unconventional gas boom unleashed in the United States.

For critics of the coal industry, Tinkler's story was a kind of parable for modern Australia, which had gorged on easy money from a China-fuelled resources boom, ignored all warning signs, and doubled and tripled its bets on coal. For old industry hands, however, his rise had almost nothing to do with coal mining: he was gambling, pure and simple. It had been a frenzy of mad, debt-fuelled speculation.

When Tinkler had bought into the controversial Ferndale project in the Hunter Valley, union boss Tony Maher had complained that the industry was in the grip of profiteers: 'These guys are buying and selling mining tenements. They're real estate transactions rather than mining. It's a Monopoly game for these people.' Keith de Lacy, chairman of Macarthur Coal, when asked to sum up Tinkler's contribution to the mining industry, put it bluntly: 'I don't think Nathan really made a contribution to the mining industry. He never developed an operating mine.'

Since he'd started out with nothing, Tinkler had to roll the dice to get ahead. Initially at least, most people had wanted Tinkler to succeed. He was the kind of guy Australians love to love. His tradie-to-billionnaire tale was an extreme validation of the aspirational politics cultivated by both major parties since the economic rationalism of the 1980s, whether it was characterised as 'Howard's battlers' or Mark Latham's 'ladder of opportunity'. In a resources boom, and amid a chronic skills shortage, vast numbers of tradesmen had become small businessmen – self-employed contractors or sub-contractors. The money was way better than in many traditional white-collar jobs, and much of it was earned in regional areas, where it was harder to spend it. Six-figure incomes allowed young blokes to pay off a house in their 20s or 30s, and even to start building a property portfolio. Forget about class warfare, there was money to be made, and the bosses and the workers were in it together.

Once Tinkler made it big, however, with hundreds of employees and suppliers across his business empire, he had turned on people, sometimes savagely. He wasn't a class traitor: he was a pure bully. When *Four Corners*' reporter Stephen Long interviewed Bruce and Tim Curry, electrical contractors near Tamworth, who had been owed $20,000 by Tinkler for almost three years, he asked Tim how it felt to have been done over by a fellow sparkie, a small businessman trying to get ahead. Tim's response was telling: 'I was pretty gobsmacked, to say the least. I thought someone like himself, on his rise to the top, he would remember where he came from.' Tinkler remembered, all right, but he couldn't care less.

A miserly attitude is not uncommon among the very rich – often enough, it's how they got there. Even vindictiveness is par for the course, and it's hardly surprising when multimillionaires and billionaires push people to the wall, just because they can. This was compounded, in Tinkler's case, by the fact that, despite his paper wealth and outward success, he was almost always short of cash, if not on the verge of going broke. Tinkler never, in his brief time at the top, had a dependable source of income. He was always buying the next thing before he had the money. He was in a perpetual cash-flow crisis, and it was entirely self-inflicted. It was also entirely selfish: money could always be found for the next horse, car or house, but never for the small creditor.

It was the harsh treatment of ordinary people that hurt Tinkler's reputation the most, particularly in the Hunter Valley. As former journalist Neil Jameson told Long, it was 'very, very damaging' for Tinkler to owe so many people money in a city like Newcastle:

> ... particularly when you're hearing that you're indulging yourself with Bentleys, Ferraris, executive jets and helicopters, and an electrician with a young family is struggling because he's owed a significant amount of money, it doesn't go down too well ... and there was a conga line of those people. The procession of debtors went from here to Muswellbrook and back. Hunter Valley people couldn't give a toss whether Nathan stiffed a hedge fund in Singapore, or a US bank, but they sure care about the fact that he might've stiffed a whole bunch of small people.

Some have compared Tinkler to a latter-day Ned Kelly – a business bushranger, hailing from nowhere and shoving it up the establishment. Of Australian-Irish stock himself, Tinkler might indeed enjoy the comparison. But the Kelly legend was built around his famous decency to the ordinary men and women he moved among and often relied upon – as when Ned and his gang made a night of it with their 'prisoners' at Euroa. Tinkler could turn on the charm, but when it came down to money, what he most often showed ordinary people was contempt.

Kelly's grievance was with the police. Kelly was persecuted; Tinkler wasn't. Kelly stole but wasn't greedy; Tinkler's greed knew no bounds. Like Kelly, Tinkler had an acute way with words. But he's produced no Jerilderie Letter as yet, just rants in old papers, takedowns of all and sundry, his words undermined by his deeds.

Others have compared Tinkler to a junior Alan Bond. But Bond's rise took two decades, and he built a business empire worth many billions that spanned the globe, from beer to media, property to mining. When the Bond Corporation collapsed, after being propped up by lazy banks and 'WA Inc.', many thousands of shareholders lost millions of dollars. Bond didn't pay his bills either, but Tinkler did not rip off thousands of investors. What money Tinkler lost was mainly his own. Over and over, he ripped himself off. Tinkler has left behind a trail of smiling vendors, as well as dozens, if not hundreds, of small creditors.

Like Bond, a love of sport may redeem Tinkler in the public mind. Bond will always be a hero for winning the America's Cup for Australia in 1983 – indeed, for many people, it is all they know about him. It is hardly the same, but if the Newcastle Knights win themselves a Grand Final, under coach Wayne Bennett, Tinkler will go down as a legend in the history of rugby league for turning a much-loved club around. This alone would be a greater achievement than many of us can point to in a lifetime.

Of course, Tinkler did a lot more than that, and his story is far from over. He has quit his Singapore palace and is said to be living in New York; the status of his marriage is unclear. Tinkler can still call on powerful friends: people such as Mark Vaile, Wayne Bennett and John Singleton – who leapt to his mate's defence in June, telling the *AFR* that Tinkler was 'a bloody good bloke … I like spending business time with him, drinking time with him and family time with him.'

Tinkler's mining empire is gone and a pile of debt hangs over his head, but Patinack is still racing, Buildev is still building, and the Hunter Sports Group is hanging on to Newcastle's two beloved football teams, at least for now. Apparently, he still flies first-class. It is hard to see how all that can continue, but Tinkler – all in, every bet – is full of surprises. As one mineworker in a Muswellbrook pub told me: 'I'd like to be as broke as he is.'

12.

EPILOGUE

'Who is icac,' Tinkler shot from his iPad. 'NSW gov,' answered Darren Williams, the young turk running Buildev, the development and construction business which Tinkler co-owned and which was about to land him at the centre of a sensational corruption inquiry. Tinkler lost it: 'Oh mate u r fucking kidding me ... what have I ever had to do with this business, can't trust anyone.'

It was 19 April 2013. A day earlier, a notice addressed to Tinkler, as director of Patinack Farm, had turned up at the group's Newcastle offices. He had a week to give the Independent Commission Against Corruption any records of work done for Patinack by a company called Eightbyfive or by one Timothy Koelma between 2009 and mid-2012.

It is hard to believe Tinkler was really unaware of ICAC in 2013: he'd met Moses Obeid over the Mount Penny project, and wound up investing alongside the Obeids when he bought a stake in the listed Coalworks. Both deals were the subject of public hearings in ICAC's Operation Jasper, underway at the time.

Tinkler and his team were likely unaware that they were being watched by ICAC, which was gathering Williams' emails, texts and mobile phone records. These were damning: from mid-2010 until early 2012, Buildev lobbied fiercely for Tinkler's proposed coal loader on the old BHP site in Newcastle, and knowingly skirted rules which prohibited developers from making political donations.

After years spent targeting corruption within the New South Wales Labor Party, ICAC's Operation Spicer was now focusing on potentially corrupt Liberal Party fundraising – particularly through Eightbyfive, a slush fund set up by Koelma, a staffer for the Liberal member for Terrigal, Chris Hartcher, who would soon become the state's energy and resources minister.

Koelma's job was party 'black ops'. He dealt exclusively with Williams, who arranged payment of some $66,000 to Eightbyfive – generally at $5000 a month, starting in June 2010.

Eightbyfive's invoices weren't paid by Buildev, however. Oddly, they were sent to Patinack, which had no need of political favours. As later emerged in ICAC public hearings, this had been the subject of emails between Williams and David Sharpe, Buildev's founder.

At 4.57 pm on Wednesday 2 June 2010, Williams asked Sharpe which entity should make the donations. Twenty minutes later, Sharpe told Williams: 'Ask Nathan as I think it's best to come through patnack (*sic*) get right away from property minning infristructure (*sic*).'

At 10.47 am the next day, Williams emailed back to check: 'Do I ring Nathan or troy?' (He was referring to Troy Palmer, Tinkler's finance chief.) A minute later, Sharpe replied: 'nt.' Phone records showed that, four minutes later, at 10.52 am, Williams did indeed call Tinkler.

When shown this email trail almost four years later, Tinkler told ICAC's fearsome assisting counsel, Geoffrey Watson SC, that the conversation he'd had with Williams could have been about anything; he had 'never spoken to Darren about making donations to the Liberal Party'.

WATSON: Well, I'm putting it to you right now that you're lying and that this call …
TINKLER: I am not …

> WATSON: ... this call was by Darren Williams, following up with you to find out the best entity to be used to disguise payments to the Liberal Party.
>
> TINKLER: No.

The verbal sparring continued, until an exasperated Watson asked whether Williams had rung Tinkler 'out of the blue [to talk] footy scores, something like that'. Tinkler played a dead bat: 'Probably footy scores, yeah.' Without an actual recording of the phone call, ICAC couldn't prove a thing.

Tinkler was a remarkable witness: he was brazen, defiant, some said smug. He cracked jokes – about how repetitive the questioning was ('Jeez, I'm starting to see why this has been going for three weeks'), about how he donated to everyone but the Greens ('I use recycled toilet paper ... that's enough for the Greens'). He answered back, and even started asking questions himself, at which point Watson exclaimed: 'Good Lord, you realise you're sitting here in a witness seat at ICAC?' During the lunch adjournment, Tinkler was overheard telling someone that the inquiry was 'some of the most boring shit I've ever seen'. Seasoned ICAC reporters in the squashy media room reckoned Tinkler was smashing it.

Tinkler's performance belied the gravity of the inquiry. He denied any knowledge of Patinack's donations to 'Eightybyfive', as he twice called it. He denied knowledge of the $5000 donations to the National Party made by Aston Resources' bosses Todd Hannigan and Tom Todd and their wives in late 2010. He was not aware that Buildev was funding the campaign of the Liberal Party's candidate for the seat of Newcastle, Tim Owen.

Tinkler did admit he'd made a personal donation to the National Party of $50,000, which was rejected – as a director and shareholder of Buildev, he was a prohibited donor. (And Tinkler never got his money back, as he pointed out.) He accepted he knew about a $53,000 donation to the Liberal Party made by Boardwalk

Resources, but said he had no idea that it had actually been paid to another fundraising vehicle, the Free Enterprise Foundation.

Was Tinkler buying political favours, Watson asked.

'I've never had a political favour in my life,' Tinkler insisted. He just wanted a hearing. He was supporting the democratic system. 'Do you think I'm the only person in business that donates to a major party?' he taunted. 'Is that what I'm on trial for?'

Tinkler spoke freely about his support – to the tune of another $50,000 – for the Newcastle Alliance, a group of like-minded businessmen who hoped to turn Newcastle, for a century safe Labor territory, into a marginal seat, so it would get what they believed to be a fairer level of state funding.

This was all very high-minded, but the texts bouncing around revealed an uglier agenda: to get rid of the sitting MP, Jodi McKay, Labor minister for the Hunter, who was opposing Buildev's proposed coal terminal. Their code for a donation was 'carpet', because the head of the alliance was Paul Murphy, owner of Churchills Carpet Court in town. On 11 March 2011, two weeks out from the state election, Williams texted Tinkler:

> WILLIAMS: You ok mate if we get some more carpet?
> TINKLER: Gees how much?
> WILLIAMS: You want her gone don't you? 50
> TINKLER: How much is sharpie puttin in? Generosity starting to get tested but yeah whatever it takes.

On the witness stand at ICAC, Tinkler made no apology for wanting McKay out of the state parliament. 'She was all about keeping Newcastle looking like Old Sydney Town,' he said. Could the Newcastle Alliance campaign have made a difference to her re-election chances? 'I hope so,' Tinkler replied.

But Tinkler denied any involvement in the real skulduggery, the letterboxing done just days out from the poll. Thousands of unauthorised colour leaflets bearing the heading 'Stop Jodi's Trucks' claimed that a container terminal – McKay's publicly preferred option for the ex-BHP site – would mean an extra 1000 truck movements a day, 365 days a year, through the surrounding suburbs. The campaign was hugely damaging to McKay, who had to visit all these homes again and explain her true position. On election day, when she ran a close second to Tim Owen, missing out by only hundreds of votes, it was clear that the leaflet had made a real difference.

McKay contemplated a legal challenge. The matter could hardly be more serious: the make-up of parliament had been altered. McKay's own evidence to ICAC showed that the leaflet was printed in Western Sydney for an 'Ann'. This was almost certainly one of her own ex-staffers, Ann Wills, who by now was working alongside mayor and independent rival candidate John Tate, and giving PR advice to Buildev. She was also the eyes and ears in Newcastle for the ex-Labor ports minister, corrupt powerbroker Joe Tripodi.

When McKay finally took the witness stand to tell her story, she surprised everyone, including herself, breaking down in tears when Watson confirmed something she had long suspected: 'We've got some pretty good information, Ms McKay, that indicates there were three people or entities behind this document, and I'll name them. Firstly, we'll call it the Tinkler Group; secondly, Ms Ann Wills; and thirdly, Mr Joe Tripodi.'

McKay was also asked about her allegation that Tinkler had offered to donate to her campaign, and that he had told her he could get around the prohibited donor laws by using his employees.

> WATSON: ... it does sound to me like a bribe, an attempt to bribe you?
> McKAY: It certainly, it certainly felt like that, he wanted my support and he was prepared to buy that.
> WATSON: Another word for that would be bribe?
> McKAY: Yes.

Ahead of his ICAC appearance, an upbeat Tinkler had told the throng of journalists he was looking forward to his turn on the stand, where he could put an end to three weeks of character assassination, and how there'd be none of McKay's 'crocodile tears'. Now, in the box, he accused McKay of lying.

WATSON: You then offered to donate to her campaign to assist her, didn't you?
TINKLER: No, I did not.
WATSON: You deny that, do you?
TINKLER: Yeah, I certainly do deny that.
WATSON: Let's make it clear?
TINKLER: Yeah, no, I wanted her gone, I was on the other side.

Tinkler also put himself at odds with Watson by denying outright that he had anything to do with the 'Stop Jodi's Trucks' leaflet, paid for by Buildev. He did not know Ann Wills. He had never met Joe Tripodi and had no idea he was working for Buildev. He had never spoken with Williams about the letterbox drop, nor had he funded it. The very suggestion was 'ridiculous'. Watson was content to let this hang.

WATSON: I'm asking you whether you've ever seen this before?
TINKLER: No, I've never, I've never seen one of these brochures before.
WATSON: Never heard of it?
TINKLER: No.
WATSON: Didn't finance it?
TINKLER: No.
WATSON: I really want to give you an opportunity … here to answer this carefully.
TINKLER: Yeah.
WATSON: Please do not … brush it aside with a laugh.
TINKLER: No, no, please, I had nothing to do with this.

Right at the end of his evidence, Tinkler's own barrister tried to turn the whole bribery episode on its head, asking whether McKay had in fact solicited a donation from Tinkler, but the insinuation was quickly withdrawn after objections from Watson and Commissioner Megan Latham. This very serious allegation had never been put to McKay.

Tinkler was excused, but he was not discharged. He was free to return to his home in Singapore, but the inquiry would be reconvening in August, he was told, and he might still be recalled. ICAC had already brought down a premier, Barry O'Farrell, police minister Chris Hartcher, and minister for the Hunter Mike Gallacher, while Newcastle MP Tim Owen had declared he would not recontest his seat given that his election campaign had been supported by 'under the table' payments from Buildev, a prohibited donor. Who knew where it would end up?

As Tinkler went to leave, a process server was waiting for him with a claim for an unpaid debt. With a heavy media contingent watching his every move, it was beyond embarrassing. Tinkler hid in a back room reserved for witnesses while his barrister accepted the documents. Later, in a lift, Tinkler's barrister told journalists the debt was 'trivial', but in fact the process server was from the international law firm Allen & Overy, which was unlikely to be acting for some piddling client.

By the time he reached the ground floor foyer, Tinkler's mood was dark. A full media pack pressed up to the sliding doors, with lights, cameras and mics jostling. Tinkler squared up, set his face and barged through the scrum up the road to his lawyer's office, tripping the swarming hacks over backwards. There was a bit of mongrel left in the old footy player.

*

Suddenly, Tinkler was front-page news again. For six months, there had been only one story: the slow, predictable wind-up of the former billionaire's empire, as he was stripped of asset after asset. Now, his world was turning inside out.

Tinkler's three-quarter stake in the valuable Middlemount royalty was taken from him by his lender, Jefferies, in November 2013. Tinkler's people said the royalty was sold at a profit – not hard, given he'd paid nothing for it – but they kept the price secret. In fact, it wasn't a sale at all: Jefferies was exercising its security for the US$24-million loan it had extended Tinkler a year earlier.

Soon he would sell off the adjoining blocks at Ocean Street, Merewether, also secured to Jefferies, where he and Rebecca had planned Newcastle's most expensive mansion. The *Daily Telegraph* snapped Rebecca at Sydney airport, and Tinkler's spokesman confirmed the long-rumoured split: she was now living in Maui with the kids.

Receivers had taken over Aston Metals, Tinkler's last substantial mining asset; by early 2014 it had been bought by one of his former chiefs, Hamish Collins, who by now was running his own listed exploration company. Tinkler didn't get a cent from the galling transaction: all proceeds went to secured lenders.

Inevitably, painfully, Tinkler lost control of the Newcastle Knights. Despite being given a two-month extension, he failed to top up a $10.5-million bank guarantee that was due by the end of March. After years of rumour, Westpac took the extraordinary step of confirming publicly that the funds were indeed being held on deposit, and it stood ready to make them available to support the club, if called on by the members. The members' club, chaired by solicitor Nick Dan, now had the right to buy the team back for a dollar.

Nobody wanted legal action. Tortuous negotiations between the Hunter Sports Group, the members' club and the NRL aimed at forming some kind of joint venture. But the talks soured amid reports that, after three years without a major sponsor, the Knights had accumulated losses of $20 million, including a hefty unpaid tax bill. If that was true, the club was now in a worse position than when Tinkler had taken it over.

Tinkler argued that the bank guarantee should be applied to these liabilities. The members and the NRL stonewalled: the liabilities were Tinkler's, while the guarantee was meant to provide for the Knights' future. Finally, Tinkler simply refused to put any more money into the club. On Friday 16 May 2014, the evening of Tinkler's ICAC appearance, as the Rabbitohs kicked off against the Storm, Channel Nine commentator Ray Warren made a stunning on-air announcement: at least 20 Knights players and staff had not been paid their salaries. The NRL had condemned Tinkler's behaviour as 'completely unacceptable'.

This caused a minor furore, and over the next few days the media speculated that supercoach Wayne Bennett and star fullback Darius Boyd were among those owed money. Worse, there were reports that Tinkler was sending abusive texts to Knights players, including the out-of-form Boyd. Fans and commentators alike turned hostile: Tinkler's time was up.

On 24 May Tinkler announced his exit. He talked up the achievements of his three years of ownership, but few were buying it. Good riddance, many said. Tinkler had not lived up to his promises. Bennett was not expected to stay on. After a terrible start to 2014, including a shocking spinal injury to forward Alex McKinnon, the Knights would go back to basics, becoming once again a grassroots club that aimed to nurture local talent. At a press conference in Newcastle on Saturday 14 June, the NRL finally confirmed that it would take over a majority stake in the Knights, with the members owning 20 per cent. The code's chief, Dave Smith, hosed down concerns that this amounted to a Sydney takeover: all six directors of the club would be locals. The club's finances were underpinned by half the guarantee funds, with the other half used to pay off suppliers and employees. Most pundits expected the NRL would soon on-sell the club to the local pokies kings, Wests, which owned the Knights' training and admin facilities at Mayfield.

The deal left the Hunter Sports Group, rumoured still to have a hefty tax liability, in a precarious position, and there was plenty of speculation that the Newcastle Jets was on the market and would soon be sold. The announcement ended a two-year stand-off between the Knights and their local paper. For the *Newcastle Herald*, it was an utter vindication.

A week after his ICAC appearance, Tinkler finally announced the sale of Patinack Farm, the stud he'd dreamed of owning since childhood. On the market for almost two years, Patinack was now down to four stallions and 600 horses at its two farms: Sandy Hollow in the Hunter Valley, and Canungra in the Gold Coast hinterland. The sale price wasn't revealed but media reports put it north of $100 million.

No doubt Patinack still had valuable assets, but how much debt was there? The deal was supposedly done on a 'walk-in, walk-out' basis – which is to say the buyer would take over the property, the bloodstock and the workers, holus-bolus. Most curiously, the sale was to a mysterious consortium of unidentified local and Middle Eastern investors, and had been negotiated by a United Arab Emirates-based firm called Cibola Capital. This company had a ludicrous website set up just nine months earlier, and was headed by an Australian builder, Daniel Kenny, who appeared to exist only in the Patinack press release. No one in the racing industry had ever heard of Kenny, and journalists' queries to Cibola went unanswered. The whole deal looked suspicious, and creditors were still circling.

In the middle of this blizzard of negative publicity, when things could hardly have looked worse for Tinkler, the US coal giant

Peabody Energy made a brief, stunning announcement to the New York Stock Exchange. Its mothballed Wilkie Creek mine, in Queensland's Surat Basin, had been sold to Bentley Resources, the Singapore shelf company Tinkler had set up in 2012, for $70 million in cash plus assumption of liabilities. The deal wasn't yet closed, being subject to certain unspecified conditions. Tinkler was back in the game!

The story was dropped to the *Australian*, which put the value of the deal at $150 million. It wasn't clear how much was debt, but Jefferies was in there. Jefferies had merged with Leucadia, the US private equiteer which had made a fortune out of Australian billionaire Andrew Forrest's Fortescue Metals, lending him $100 million on advantageous terms when he was in a tight spot. Perhaps they were about to do the same for Tinkler?

Tinkler told the *Australian* that despite the very tough world coal market, it was a good time to buy: 'We're in a value part of the cycle.' He reckoned he'd learned 'a lot of lessons' from his character-building Whitehaven experience. 'I have my hands firmly on the wheel this time,' he said, 'and I'm not letting go.'

The cynics had a field day. The closed mine, which had been on the market for two years, was literally worthless. Like Tinkler's other deals, this one would be totally debt-funded – at exorbitant interest rates and on sore terms – with Tinkler putting down a bare minimum of his own cash and scrambling to raise the rest of the purchase price. It was the perfect way to lose money. And didn't he still owe millions to Farallon's Ray Zage?

Who cared – that was Farallon's problem. It might be a one-in-a-million chance, but Tinkler was on the comeback trail. The glamour was gone, the boganaire trappings all stripped away. The 'pit leco' was getting back to his core business. If he was going to go down, he would go down fighting.

ACKNOWLEDGEMENTS

A biography of Nathan Tinkler was the very good idea of my friend and then colleague Matt O'Sullivan, who suggested it over a beer at the Point Hotel in Pyrmont as we watched the 2012 Melbourne Cup. I had just been injuncted by Tinkler and was having a good old whinge about everything, including the lacklustre sales of my book on the coal seam gas industry. Matt said: 'Nathan Tinkler – that's what people want to read about.'

It was immediately obvious to me that he was right, and his idea remained with me throughout the three-month court case, the lifting of the injunction and my tumultuous departure from Fairfax Media. Suddenly and unexpectedly, I had time on my hands. Thus, I have Fairfax to thank, in more ways than one, for giving me the chance to write this book, and I record my appreciation for the company's support during an expensive and stressful court case, even if its outcome was not what we hoped for: the right to publish. In particular, I offer my thanks to the then editor-in-chief of the *Sydney Morning Herald*, Sean Aylmer, business editor Eric Johnston, and Minter Ellison's Peter Bartlett.

I also record my deep appreciation to Chris Feik of Black Inc. for getting in touch soon after I left Fairfax, when I was not feeling my absolute strongest, and for having the confidence to support and publish the book. My editor at Black Inc., Julian Welch, was a pillar of calm, strength and wisdom. Thanks also, in no particular order, to Anna Lensky, Imogen Kandel and Erik Jensen.

The executive producer of the ABC's *Four Corners*, Sue Spencer, provided me with a fabulous experience as I worked with the incomparable team on Australian television's finest current affairs program to produce 'In Search of Nathan Tinkler', and gave me access to invaluable transcripts and other research material that my meagre advance would never have covered. Reporter and friend Stephen Long uncovered some of the best stories that have found their way into this book, and made very helpful suggestions on a first draft of the manuscript. Thanks also to producer Clay Hichens, who stuck with the story despite the withering reluctance of people to go on camera, and researcher Mario Christodoulou, who (among many other things) did the investigation into Tinkler's coal loader proposal, and gave me a folder of documents invaluable in that section. I can't thank you all enough.

I would also like to acknowledge the many excellent fellow journalists who have had a crack at Tinkler's story, and whose work I have relied upon. At the *Newcastle Herald*, an heroic paper, Neil Jameson, Donna Page, Ian Kirkwood and Robert Dillon all did major pieces on Tinkler. At the *Australian Financial Review*, the investigations and features of Neil Chenoweth, Angus Grigg, Jamie Freed, Hannah Low and Anne Hyland were my constant companions. At *Good Weekend*, Ben Hills wrote a cracking profile of Tinkler, a touchstone for this book, and helped me with some key contacts. At the *Sydney Morning Herald*, Tom Reilly – the first to really probe Tinkler's darker side, whose story first carried the book's title, *Boganaire* – read part of the manuscript and offered suggestions; I've also drawn freely on the work of my former *SMH* colleagues Stuart Washington, Ben Butler, Kate Lahey, Chris Roots and Michael Cockerill. At *BRW* (and formerly Smart Company), James Thomson and John Stensholt wrote consistently insightful pieces on Tinkler. At the ABC's *Media Watch*, Paul Barry, who wrote significant stories on Tinkler for the *Sunday Telegraph* and for *Crikey*, gave me encouragement and a tip or two. I have also quoted from the work of Brendan King at the ABC's *Background Briefing*, and of Conor Duffy at the ABC's *7.30*. Beyond these, there have been many, many more great stories published or broadcast about Tinkler – quite often under a cloud of threats and intimidation – than I have the time or space to list.

Thanks also to Steve Normoyle at Chevron Publishing, John Shakespeare at the *Sydney Morning Herald*, and John Hayes at Correct Planning and Consultation for Mayfield. My former News Limited colleagues Vanda Carson and Anthony Klan were great company at the interminable Tinkler court hearings in 2012 and 2013.

Special thanks to my new editors at *Crikey*, Jason Whittaker and Cathy Alexander, and to publishers Eric Beecher and Marina Go, for waiting for me.

Thanks to Jan McLelland and Tom Morton at the Australian Centre for Independent Journalism, at the University of Technology, Sydney, for appointing me as an honorary associate of the centre, thus enabling me to use the library's databases. At the New South Wales Supreme Court's media unit, Jo Oakes and Sonya Zadel put up with my repeated requests for the many legal files concerning Tinkler. Thanks to Matthew Abbott and Andre Khoury at the Australian Securities and Investments Commission. Thanks to Karmen Herriman of Right to Information Services at the Queensland Department of Natural Resources and Mines. Thanks also to Kerrie Hawkins and Donna at the *Muswellbrook Chronicle*, for helping me access and copy clips from the archive.

Thanks to my father, Peter Manning, for help with the Tinkler family tree. Thanks to my wife, Melinda, for everything. Thanks to Mum, Henry and Maureen for all the help with the kids. Thanks to my awesome boys, Jude and Milo, for being so awesome. Thanks to my dear friend Roberta Ivers, for support and advice. Thanks to my fantastic neighbours.

Finally, to all the people who helped anonymously or off the record – you know who you are – and particularly to those who saw some public interest in revealing the truth about Tinkler, and those who read and commented on drafts of the manuscript: this book is for you.

NOTES

This book is based on countless interviews, both on and off the record, private correspondence, press clippings, announcements, company filings, title searches, court filings, government records and some confidential documentation which I can neither acknowledge nor attribute. Transcripts of emails and text messages are reproduced verbatim. As far as possible, I have tried to acknowledge critical sources in the body of the text. What follows is therefore selective.

I relied on some key general sources throughout the book, and I recommend them to anyone interested in reading further on Tinkler. In rough chronological order, they are:

- *BRW*'s Rich 200 (May) and Young Rich lists (October). For the years 2008–13 these lists chronicle Tinkler's rise and fall. Their estimates of Tinkler's wealth are referred to throughout this book, despite my reservations about whether the methodology at all times appraised his net wealth.

- Neil Jameson's 'Big Spender' (Newcastle Herald, 31 January 2009) was one of the first profiles of Tinkler, and remains one of the best.

- In September 2010 Tinkler did a vital on-the-record interview with Angus Grigg of the *Australian Financial Review*. Grigg's published pieces from this interview included 'The Deal Maker' (with Neil Chenoweth, *BRW*, 30 September 2010), 'From School Failure to a Class of His Own' (with Neil Chenoweth, *AFR*, 13 October 2010) and 'Tinkler bets on Revenge' (*AFR*, 16 October 2010). I am extremely grateful to Grigg for giving me access to the unpublished transcript of that interview.

Tom Reilly's 'Mining Boss Rocks Racing' (*Sydney Morning Herald*, 3 April 2010) was the first article to suggest Nathan Tinkler was not paying his bills. Reilly's 'Living like a Boganaire' (*SMH*, 10 October 2010) provided the title for this book. A string of Reilly's other articles are referenced throughout the book, especially on Patinack.

Sarah-Jane Tasker, 'Business Giant with Humble Aims', *The Australian*, 13 December 2010.

Nathan Tinkler's interview with Mike Rabbitt of NBN News on 4 March 2011 is available on YouTube.

Brendan King, 'The Knight of Newcastle', *Background Briefing*, ABC Radio National, 19 June 2011.

Angus Grigg and Jamie Freed, 'Tinkler Puts His Empire on the Line', *AFR*, 10 December 2011.

Elisabeth Behrmann, 'Laborer-to-Billionaire Tinkler Plans More Coal Bids After Whitehaven Deal', *Bloomberg*, 13 December 2011.

Tim Treadgold, 'Country Boy Makes it Good in Coal', *Forbes*, 1 February 2012.

Donna Page, 'Trail of Debt', *Newcastle Herald*, 20 August 2012.

Ben Hills' 'Boom and Bust' (*Good Weekend*, 15 September 2012) is a brilliant profile of Nathan Tinkler, and was a touchstone for this book.

Conor Duffy, 'Australian Billionaire's Rise Faces Fall from Creditors Call', *7.30*, ABC Television, 23 October 2012.

Hannah Low, 'A Price to Pay', *AFR*, 27–28 July 2013.

Stephen Long, 'In Search of Nathan Tinkler', *Four Corners*, ABC Television, 29 July 2013. As noted elsewhere, Stephen provided me with many of the key anecdotes, interview notes and transcripts used in this book, many of which are not publicly available. Once again, I am grateful to Stephen, Mario Christodolou, Clay Hichens and Sue Spencer for use of this material.

Key chapter-specific sources follow. Where published articles are cited, it is because they are mentioned in the text or quoted at length.

CHAPTER 1: PIT LECO – SCHOOL OF HARD KNOCKS

Quotes from Nathan Tinkler in this chapter are from Angus Grigg's interview, unless stated otherwise.

A copy of Norman Tinkler's obituary in the *Port Macquarie News*, 3 August 1979, is held at the State Library of New South Wales. Most of the official records for Nathan Tinkler's life – his birth and marriage certificates, for example – are not yet publicly available because of privacy laws. Similarly, records of Tinkler's attendance at or graduation from primary and secondary school, and Muswellbrook TAFE, are not publicly available.

Quotes from members of the Muswellbrook Rams are from *Four Corners* transcripts.

'The Interview', *Master Electrician*, Winter 2010, pp. 41–43.

Recollections of Tinkler's time at Mount Arthur and Bengalla are based on private conversations with former colleagues.

Wayne Kempster's story is based on interviews and correspondence.

The section dealing with Tinkler's employment with John Moore and DTMS is based on interviews, correspondence, and filings and consultations with ASIC.

Matthew Higgins' account of his first meeting with Tinkler, and of his employment at DTMS, is based on his affidavit *In the Matter of Oceltip Pty Ltd*, case number 11507, dated 12 December 2012, Supreme Court of Queensland.

A number of orders against Tinkler were obtained from the Singleton Local Court.

Perpetual Trustees v Nathan Tinkler, Supreme Court of New South Wales, case number 11349 of 2004.

Ian Kirkwood, of the *Newcastle Herald*, pointed me to the phrase 'profitless prosperity'. See also his book *Newcastle: New century, new horizons* (Focus Publishing, 2000, with Christopher Ford) for background to this section and the introduction of Chapter 6, below.

Geoff Evans, 'A Just Transition to Sustainability in a Climate Change Hot Spot: The Hunter Valley, Australia', an unpublished PhD thesis at the University of Newcastle, September 2009, provided some useful background to this section on coal markets and the Hunter Valley.

The *Muswellbrook Chronicle*'s archive for the years 1998–2002 is not available online, but is available at the newspaper's office in Muswellbrook.

CHAPTER 2: MIDDLEMOUNT – FROM $1 MILLION TO $442 MILLION

The sale of Tinkler's labour hire business was not reported at the time. This brief account is based on private conversations with sources close to the transaction.

My account of Peter Mallios' involvement with Tinkler is based on affidavits in *Peter Mallios v Custom Mining Ltd*, Supreme Court of New South Wales, case number 50109 of 2008.

This account of the sale of Middlemount by DJB and Sennen to Custom Mining, and of its exploration and development, is based partly on a Right to Information request to the Queensland government for all records concerning the tenement MDL 282. Through that and other means, I have obtained copies of reports including: Information Memorandum, Middlemount Coal Deposit, Ernst & Young, July 2005; Annual reports on MDL 282 for 2002–03, 2003–04, 2004–05, 2005–06 and 2006–07; Custom Mining Middlemount's environmental management plan bulk sampling program June 2007; and Middlemount Coal Project – Custom Mining, comments by Ray Smith, 21 June 2007.

Richard Jennings' story is documented in the ASIC filings, including the minutes of a meeting of Custom Mining Ltd on 7 October 2006.

The Middlemount sale agreements between DJB and Sennen and Custom Mining Ltd were not released, but the structure of the deal, and the deferred payments, can be pieced together from Sennen's announcements to the Toronto Stock Exchange, and from the 2006–07 accounts of Custom Mining Ltd, available from ASIC. Those accounts also partially document Noble's investment in Middlemount, which was not announced to the Singapore Stock Exchange. The rest has been filled in from private conversations and correspondence.

The returns to one Martin Place investor in Custom Mining are exact and are based on a private conversation.

The account of the sale of Custom Mining Ltd to Macarthur Coal is based partly on private conversations with sources close to the deal; former Macarthur Coal chief Nicole Hollows' comment was made at a 'Game Changers' seminar hosted by the Queensland University of Technology Business School, at the State Library, Brisbane on 29 May 2013.

Quotes from former Macarthur Coal chairman Keith de Lacy are from the *Four Corners* transcripts.

CHAPTER 3: ALL TOO HARD? THE STORY OF PATINACK FARM

Brian Russell's 'Nathan Tinkler and Ray Bowcock: Both trailblazers in their era' (www.justracing.com.au, 31 July 2008) provided useful background information for this chapter.

Affidavits in *Galogo Bloodstock v Nathan Tinkler*, Sydney Local Court, case number 13167 of 2004; *Luxbet v Nathan Tinkler*, Supreme Court of the Northern Territory, case number 89 of 2012; *Something Fast v Patinack Farm*, Supreme Court of New South Wales, case number 291427 of 2009; and *Peter James v Nathan Tinkler*, case number 106978 of 2010.

Figures on horses bought by Patinack Farm are from www.stallions.com.au; figures on coverings by Patinack stallions are from the *Australian Stud Book* (www.studbook.org.au); figures on John Thompson's all-season statistics are from www.racingandsports.com.au. The website www.breedingracing.com.au was also a very helpful source.

Les Tinkler was interviewed by Pat McLeod: 'Tinkler Racing's Man of the Moment', *Gold Coast Bulletin*, 27 March 2008.

The passage quoted at length is from Tom Reilly, 'Big Bills, Few Winners, for Tinkler the Breeding Tyro', *Sydney Morning Herald*, 29 October 2011. The unpublished quote was relayed to me by Reilly.

Tom Reilly, 'Horse Trading over Pasta, Wine and Legal Agreement', *Sydney Morning Herald*, 14 October 2011.

CHAPTER 4: SPENDATHON – BOYS AND THEIR TOYS

Property data is from New South Wales Land and Property Information records, press clippings and www.onthehouse.com.au.

Aircraft registration information is from www.airframes.com; descriptions of Tinkler's planes come from sites including www.controller.com.

Tim Sommers' ebook *The Supercar Club* (Wabow, 2011) was available free online at http://dev.hoth.net.au/misc/SuperCar-eBook.pdf at the time of publishing.

This account is based partly on the judgement and affidavits in *Supercar International Holdings v Sommers; Tinkler Group Holdings v Sommers*, Supreme Court of New South Wales, case number 4304 of 2009.

Paul Barry, 'Boys Played with Luxury Toys and Lost', *Sunday Telegraph*, 13 June 2010.

Ros Reines, 'Hush' column, *Sunday Telegraph*, 28 September 2008.

The account of Tinkler's surgery is based on private conversations; see also the Confederation of Australian Motor Sports' Medical Standards for Fitness to Compete (August 2009) at www.cams.com.au.

Dick Johnson and James Phelps, *Dick Johnson: The Autobiography of a True-Blue Aussie Sporting Legend*, Ebury Australia, 2013.

CHAPTER 5: MAULES CREEK – LIGHTNING STRIKES TWICE

This chapter is largely based on press clippings, company announcements and filings, government records and private interviews and correspondence with the parties involved.

Stuart Washington's 'How Singo Made Tinkler Rich' (*Sydney Morning Herald*, 30 October 2011) uncovered the connections between Farallon, Carnegie and Singleton.

Tony Windsor's quotes are from Paddy Manning, 'Mining for Controversy', *Sydney Morning Herald*, 29 January 2013, and the estimate of the actual financial impact of the Moylan scam was in my story 'Hoax Loss Was Less than Half a Million', *Sydney Morning Herald*, 16 January 2013.

For *Northern Inland Council for the Environment v Minister for Environment* (Maules Creek and Boggabri Mine cases), see www.edo.org.au.

Political donations data is from www.aec.gov.au and www.demoracy4sale.org.au; NSW Greens MP Jeremy Buckingham's quotes are from the *Four Corners* transcripts.

CHAPTER 6: TINKLERTOWN – THE WHITE KNIGHT RIDES IN

For the comparison of 'coal barons' Tinkler, Palmer and Rinehart, see Guy Pearse, David McKnight and Bob Burton, *Big Coal: Australia's dirtiest habit* (NewSouth Books, 2013), p. 77.

Quotes from Neil Jameson, former *Newcastle Herald* journalist and Jets Advisory board member, are from the *Four Corners* transcripts.

Aaron Kearney's quotes are from Brendan King's *Background Briefing* story on ABC Radio National, cited above.

Neil Jameson, *Our Town, Our Team: The Story of the Newcastle Knights*, Ironbark Press, 1992.

Paul Crawley, 'I Want Wayne Bennett, Reveals Nathan Tinkler', *The Daily Telegraph*, 21 February 2011.

Phil Rothfield, 'Why the Newcastle Knights Took the Fight to Nathan Tinkler', *The Daily Telegraph*, 23 February 2011; this report contained Tinkler's emails to Robert Tew.

Robert Dillon, 'Game Changers: Hunter Sports Group', *Newcastle Herald*, 15 June 2012.

Michael Cockerill, 'War Is Over: Lowy, Tinkler Call Jets Truce', *Sydney Morning Herald*, 2 May 2012.

Ian Kirkwood's 'Tinkler Eyes Ex-Steel Site' (*Newcastle Herald*, 2 December 2010) broke the story of Tinkler's coal loader, and a stream of stories followed.

I am indebted to John Hayes for copies of relevant Correct Planning and Consultation for Mayfield correspondence, submissions and leaflets.

A freedom-of-information application and other groundbreaking research by Mario Christodoulou of the ABC's *Four Corners* provided crucial background for the coal loader section of this chapter. Documents obtained include the New South Wales Treasury's 'Review of Proposed Uses of Mayfield and Intertrade Lands at Newcastle Port' (4 February 2011), and parts of Buildev's commercial-in-confidence 'Discussion Document: Former BHP Site and Newcastle Port Lands'.

CHAPTER 7: WHITEHAVEN – THE BOARDWALK TRANSACTION

My account of Tinkler's meeting with Moses Obeid and Gardner Brook is based on Hannah Low and Michaela Whitbourn's 'Tinkler's Secret Obeid Rendezvous' (*AFR*, 27–28 July 2013).

For Boardwalk's investment in Ferndale alongside the Obeids, see the Independent Commission Against Corruption's report on Operation Jasper, titled 'Investigation into the Conduct of Ian Macdonald, Edward Obeid Senior, Moses Obeid and Others' (July 2013), pp. 139–142.

Ian Verrender's 'Blue-blood brawl' (*Sydney Morning Herald*, 20 May 2009) provides a great account of NAB's failed tilt for AMP and more.

Unpublished documents relied on in this section include a draft prospectus for the float of Boardwalk Resources, prepared by Morgan Stanley and dated October 2011; a draft UBS indicative term sheet for 'Project Trifecta', dated October 2011; and Rothschild's 'Project Trifecta Discussion Materials', dated November 2011. The scheme documents, including the independent experts' report, were released to the ASX.

For ASIC taking an interest in the Boardwalk transaction, see Bryan Frith, 'Change in Whitehaven Merger Arrangement Favours Aston Investors', *The Australian*, 16 March 2012.

For the proxy adviser's recommendations and the judge's comments, see Paddy Manning, 'Tinkler's Treasure Chest Lies Under the Boardwalk', *Sydney Morning Herald*, 14 April 2012.

Gavin Wendt's comments were in an interview with the author when at *Four Corners*.

The account of Whitehaven Coal's EGM in Sydney on 16 April 2012 is based on my own reporting and recording of the meeting.

Paddy Manning, 'Nathan Tinkler Spends Big on Maui Pad', *Sydney Morning Herald*, 29 August 2012.

CHAPTER 8: JUDGEMENT DAYS – LET'S BURN A DEBT

Matthew Kelly, 'King of Coal Told "No"', *Newcastle Herald*, 28 January 2012.

Stuart Washington, 'Singapore Sling for Magnate Tinkler', *Sydney Morning Herald*, 9 June 2012.

Matthew Stevens, 'True Believers Raise Coal Stakes', *AFR*, 8 May 2012.

For the former mining minister Martin Ferguson's comments that the resources boom was over, see Alexandra Kirk, 'Ferguson – Olympic Dam Expansion May Still Go Ahead', *AM*, ABC Radio, 23 August 2012 (www.abc.net.au/am/content/2012/s3573908.htm).

On Tinkler's attempt to sell Patinack, see Tom Reilly and Paddy Manning, 'Has

Nathan Tinkler's Luck Run Out?', *Sydney Morning Herald*, 17 August 2012.

Tom Reilly, 'Billionaire Faces "Mutiny" over Failure to Pay Employee Super Contributions', *Sydney Morning Herald*, 20 August 2012.

Betting records are from affidavits in *Luxbet v Nathan Tinkler*, Supreme Court of the Northern Territory, case number 89 of 2012.

Most of the exchanges concerning the Blackwood placement are drawn from the affidavit of liquidator Robyn Duggan, *In the Matter of Mulsanne Resources*, case number 296966 of 2012 in the Supreme Court of New South Wales; see also the liquidator's reports to Mulsanne's creditors.

Most of the small creditors discussed here were interviewed by Donna Page for her 'Trail of Debt' story, cited above. Ensuing articles include: 'Tinkler Pays Up', *Newcastle Herald*, 22 August 2012; 'Doctors Chase Knights and Jets over Debts', *Newcastle Herald*, 1 September 2012; 'Knights Tax Debt Row', *Newcastle Herald*, 11 October 2012; and 'Action on Tinkler over Stadium Debt', *Newcastle Herald*, 7 December 2012.

CHAPTER 9: DOWNFALL – RUN FOR NATHAN

Tinkler's tell-all interviews were given in Nabila Ahmed and Jamie Freed, 'Tinkler Comes Out All Guns Blazing', *AFR*, 1 November 2012, and Ray Thomas, 'Millionaire Owner Nathan Tinkler Prepared to Quit Racing, Will Stick by Newcastle Knights and Jets', *Daily Telegraph*, 1 November 2012.

The reporting of the Whitehaven AGM is my own, and the exchanges between Stephen Mayne and Philip Christensen were published in Ben Butler's 'CBD' column, 'Tinkler Haunting Presence for Whitehaven' (*Sydney Morning Herald*, 2 November 2012).

John Thompson's comments on Sky Sports Radio were widely reported, e.g. 'Trainer Tells of Tinkler Troubles', *AAP*, 5 November 2012, and Tom Reilly with Paddy Manning, 'Staff Turn as Tinkler's Cash Woes Hit Stables', *Sydney Morning Herald*, 6 November 2012.

Phil Jacob, 'Nathan Tinkler Flies in the Face of his Creditor', *Daily Telegraph*, 17 November 2012; see also Jennifer Sexton, 'Nathan Tinkler Wants $16m Jet Returned', *Sunday Telegraph*, 6 January 2013.

Paddy Manning, '$2.7 Million Debt: Tax Office Moves to Wind Up Tinkler's Teams', *Sydney Morning Herald*, 13 December 2012.

The dispute between Tinkler and Matthew Higgins was revealed by Donna Page, in 'Friends to Foes' (*Newcastle Herald*, 13 December 2012); the details of the story, including correspondence, anecdotes and the forensic accountant's report, are all drawn from the affidavits of Higgins and his lawyer David Schwarz in *Oceltip*, cited above.

Elisabeth Behrmann, Brett Foley and Chanyaporn Chanjaroen, 'Tinkler Creditors Said to Consider Seeking Whitehaven Stake', *Bloomberg*, 28 November 2012.

The judgements and transcripts of *Aston Resources Investments and Boardwalk Resources Investments and Nathan Leslie Tinkler v Fairfax Media* (case number 6761 of 2012 in the Supreme Court of Victoria) and *Fairfax Media Limited v Aston Resources Investments and Boardwalk Resources Investments* (case number 10 of 2103 in the Court of Appeal) are restricted; but see reports of the injunction by Hannah Low, 'Tinkler Blamed Ex-Employee for Leak', *AFR*, 21 March 2013, and Ben Butler and Lucy Battersby, 'Tinkler's Super-Injunction on Fairfax Lifted', *Sydney Morning Herald*, 20 March 2013.

See 'Journalists Facing Prison for Protecting Sources', Media, Entertainment and Arts Alliance, 28 March 2013 (www.alliance.org.au/journalists-facing-prison-for-protecting-sources), and 'Uniform Shield Law Needed to Protect Confidential Sources', Media, Entertainment and Arts Alliance, 2 April 2013 (www.alliance.org.au/uniform-shield-law-needed-to-protect-confidential-sources).

On the loss of All Too Hard, see Chris Roots, 'Hard Bargain: Tinkler Sells Star Colt in $30 Million Stud Deal', *Newcastle Herald*, 14 December 2012, and James Thomson, 'Tinkler's Kingdom for a Horse: Racing Empire Sent to Knackery', *Crikey*, 17 December 2012.

CHAPTER 10: MULSANNE – BILLIONAIRE IN THE DOCK

This chapter is based largely on my own reporting and on uncorrected transcripts of evidence given in the Mulsanne liquidator's examinations in the Supreme Court of New South Wales, including by Nathan Tinkler, on 8 March 2013, 14 March 2013 and 15 March 2013.

ABC reporter Conor Duffy's questions to Nathan Tinkler are taken from the *Four Corners* transcripts.

CHAPTER 11: DEMOLITION MAN – THE BRIEFEST BILLIONAIRE

Paddy Manning, 'Tinkler Marriage Split Rumours', *Sydney Morning Herald*, 24 March 2013.

Quotes from Nathan Tinkler are taken from Nathan Exelby, 'Patinack Farm Owner Nathan Tinkler Berates Racing Industry during Public Address at Doomben', *Courier-Mail*, 27 April 2013.

James Thomson, 'Five Reasons Nathan Tinkler Might Not Be Finished Yet', *BRW*, 20 June 2013.

Gerry Harvey's quotes given to ABC News are from the *Four Corners* script or transcripts, as are quotes from Keith de Lacy, Tim Curry and Neil Jameson. Tony Maher's quotes are from Ian Kirkwood, 'Tinkler's $25m Coalworks Deal', *Newcastle Herald*, 6 December 2010.

I owe Richard Denniss of the Australia Institute for the observation that Tinkler's story – doubling and tripling up on coal – is a kind of parable for Australia in the resources boom.

CHAPTER 12: EPILOGUE

At the time of writing, transcripts and exhibits from the Independent Commission Against Corruption's Operation Spicer investigation were available online at www.icac.nsw.gov.au/investigations/current-investigations/investigationdetail/203. Jodi McKay appeared on the afternoon of Thursday 1 May 2014. Her statement is Exhibit S44. A copy of the letterboxed leaflet 'Stop Jodi's Trucks' is Exhibit S45. Nathan Tinkler appeared at the commission in the morning and afternoon sessions of Friday 16 May 2014. Texts between he and Darren Williams are contained in Exhibits S9 and S10, along with records of various political donations and invoices to Patinack Farm issued by Eightbyfive. Exhibit S48 contains the June 2010 emails between Williams and David Sharpe. For Tim Owen's campaign funding, see Michelle Harris, 'ICAC: Payments "made under the table" to Newcastle MP Tim Owen's Campaign' (*Newcastle Herald*, 5 May 2014).

For the sale of the Middlemount royalty, see Paddy Manning, 'Tinkler's Fall Continues with Loss of Key Mining Asset' (*Crikey.com.au*, 26 November 2013).

For the sale of Aston Metals, see 'Nathan Tinkler Foe Hamish Collins Enjoys the Last Laugh' (*Daily Telegraph*, 1 April 2014).

For the sale of Tinkler's Ocean Street properties, see Donna Page, 'Tinkler Cashes in on House Sale' (*Newcastle Herald*, 28 March 2014).

On Tinkler's separation, see Phil Jacob, 'Nathan Tinkler's Wife Moves Four Children to Hawaii after "Split" from Under-pressure Former Billionaire' (*Daily Telegraph*, 5 November 2013).

The sale of Patinack Farm was widely reported. For speculation about the mystery buyer, see Christian Nicolussi, 'At the Track: Patinack Farm Sale Mystery' (*Daily Telegraph*, 7 June 2014).

Similarly, the Knights saga was covered extensively, particularly by Robert Dillon of the *Newcastle Herald*, whose report 'Bank Guarantee Doubt Clouds Knights Ownership' (16 March 2014) revealed exclusively that Tinkler had defaulted on his ownership obligation. On the bank guarantee, see Jennifer Sexton, 'Nathan Tinkler's Knights Bank Guarantee Will Be Honoured: Westpac' (*Daily Telegraph*, 21 March 2014). On Tinkler's texts to players, see James Hooper, 'Knights Players on Receiving End of Abusive Texts from Nathan Tinkler's Phone' and Matthew Johns, 'Get Out Nathan Tinkler, You're No Longer Welcome at Newcastle Knights' (both *Daily Telegraph*, 23 May 2014).

On Tinkler's coal comeback, see Sarah-Jane Tasker, '$150m Mine Buy Puts Nathan Tinkler Back in Coal' (*Australian*, 14 May 2014). See also John Stensholt, 'Former Rich Lister Nathan Tinkler "Goes All in" on New Coal Venture' (*Australian Financial Review*, 15 May 2014); Jeremy Grant, 'Nathan Tinkler Makes Coal-mining Comeback with $150m Purchase' (*Financial Times*, 13 May 2014); and Paddy Manning, 'Comeback Kid: Can Tinkler Revive a Dead Mine – and His Own Fortunes?' (*Crikey.com.au*, 20 May 2014).

www.ingramcontent.com/pod-product-compliance
Lightning Source LLC
Chambersburg PA
CBHW070608170426
43200CB00012B/2618